CONCISE NANOCHEMISTRY

BY

BANKOLE-OJO OLUFUNSHO SAMUEL

A comprehensive summary of nanochemistry

Foreword

I am writing this letter to highly recommend Samuel for his outstanding work as a nanoscientist in the field of nanochemistry. I have had the pleasure of working with Samuel for the past few years and have been continually impressed by his dedication, expertise, and passion for this field.

Samuel has a deep understanding of the principles of nanochemistry and has been able to apply this knowledge to develop new materials and applications. His research has focused on the synthesis and characterization of nanomaterials for various applications, such as catalysis, energy storage, and biomedicine. His work has resulted in several publications in high-impact journals, and he has been invited to speak at several national and international conferences.

One of the things that sets Samuel apart is his ability to approach problems with creativity and ingenuity. He is constantly thinking outside the box and coming up with innovative solutions to complex challenges. His approach has led to several breakthroughs in the field of nanochemistry, and his work has been recognized by their peers.

Samuel has also demonstrated strong leadership skills, as evidenced by his ability to effectively manage his research team and collaborate with colleagues from different disciplines. He is always willing to share his knowledge and expertise with others and have been instrumental in training and mentoring junior scientists.

In addition to his scientific accomplishments, Samuel has also shown a strong commitment to community service and outreach. He had volunteered his time to several educational programs aimed at promoting science education in underserved communities. His dedication to improving the lives of others is truly admirable.

Overall, I am confident that Samuel will continue to make significant contributions to the field of nanochemistry. His dedication, expertise, and passion will continue to serve the scientific community. I highly recommend this book, "Concise Nanochemistry" without reservation.

Sincerely,

Dr. Varaprasad S.

Preface

Nanotechnology has emerged as one of the most exciting and rapidly evolving fields of science and technology in recent decades. The ability to control and manipulate matter at the nanoscale has led to the development of new materials, devices, and applications with unprecedented properties and functionalities. Nanoscience and nanotechnology are interdisciplinary fields that bring together concepts and techniques from physics, chemistry, materials science, biology, and engineering to understand and exploit the unique properties of nanomaterials.

This book, titled "Concise Nanochemistry," aims to provide a comprehensive overview of the fundamental principles, synthesis methods, and applications of nanomaterials in a concise and accessible format. The book is intended for students, researchers, and professionals who are interested in learning about the chemistry of nanomaterials and their potential applications in various fields.

The book is organized into twenty-one chapters that cover a wide range of topics related to nanochemistry. The first chapter provides an introduction to the basic concepts and properties of nanomaterials, including size and shape effects, quantum confinement, and surface properties. The third chapter discusses the various synthetic approaches for preparing nanomaterials, including top-down and bottom-up methods, as well as physical and chemical methods.

The fourth chapter focuses on the characterization techniques used to study the structure, composition, and properties of nanomaterials, including spectroscopy, microscopy, and diffraction methods. The fifth to ninth chapters discuss the properties and applications of various types of nanomaterials, such as nanoparticles, nanowires, nanotubes, and graphene.

The eleventh chapter focuses on the use of nanomaterials in energy conversion and storage, including solar cells, batteries, and fuel cells. The twelfth chapter discusses the applications of nanomaterials in biomedical fields, such as drug delivery, imaging, and tissue engineering. The thirteenth chapter provides an overview of the principles and applications of nanoelectronics, including the fabrication and properties of nanoscale electronic devices, such as transistors and sensors.

Each chapter of the book is expertly written in the field and provides a concise and up-to-date overview of the topic, including key concepts, recent advances, and future directions. The book is also richly illustrated with figures and diagrams that help to clarify the concepts and enhance the reader's understanding.

In summary, "Concise Nanochemistry" provides a comprehensive overview of the fundamental principles, synthesis methods, and applications of nanomaterials in a concise and accessible format. The book is an essential resource for students, researchers, and professionals who are interested in learning about the chemistry of nanomaterials and their potential applications in various fields.

CHAPTER ONE

1.1 Introduction to Nanochemistry

Nanochemistry is a rapidly growing field of science that focuses on the study of chemical interactions and phenomena at the nanoscale level. The field deals with the synthesis, characterization, and manipulation of nanoscale materials such as nanoparticles, nanotubes, and nanowires. Nanochemistry has emerged as a multidisciplinary field that draws on chemistry, physics, materials science, and engineering to design and create materials with unique properties and functionalities. The term "nano" refers to the size range of particles and materials that are typically between 1 and 100 nanometers in diameter.

Nanomaterials have unique physical and chemical properties that make them distinct from bulk materials. For example, the high surface area-to-volume ratio of nanoparticles allows for increased reactivity, and the quantum confinement of electrons in nanoscale materials leads to unique optical and electronic properties. These properties make nanomaterials attractive for use in a wide range of applications, including electronics, energy, medicine, and catalysis.

One of the most important applications of nanomaterials is in the field of nanoelectronics. Nanoelectronics involves the design and fabrication of electronic devices and circuits at the nanoscale level. This is achieved using nanomaterials such as nanowires, nanotubes, and graphene, which have unique electronic properties due to their size and composition. These materials have been used to develop faster and more efficient computer processors and memory devices, and they hold great promise for the development of future electronic devices.

Fig 1. Nanocubes

Image Credit: Kateryna Kon / Shutterstock

Another important application of nanomaterials is in the field of energy. Nanomaterials have been developed for use in solar cells, which convert sunlight into electricity, and for use in batteries and fuel cells, which store and release energy. For example, nanoparticles can be used as catalysts to speed up chemical reactions in fuel cells, making them more efficient.

Nanomaterials also have great potential in the field of medicine. Nanoparticles have been developed that can be used for targeted drug delivery, allowing for the precise delivery of drugs to specific cells or tissues in the body. This approach has the potential to reduce side effects associated with traditional chemotherapy, as well as increase the efficacy of treatments. Nanomaterials have also been developed for use in imaging and diagnostics, allowing for earlier and more accurate detection of diseases such as cancer.

In addition to these applications, nanomaterials have been developed for use in catalysis, which is the process of accelerating chemical reactions. These materials can be used to speed up reactions or make them more selective, leading to the development of new and more efficient chemical processes. Nanomaterials have also been used in the development of sensors, which can detect and measure the presence of specific chemicals or biological molecules.

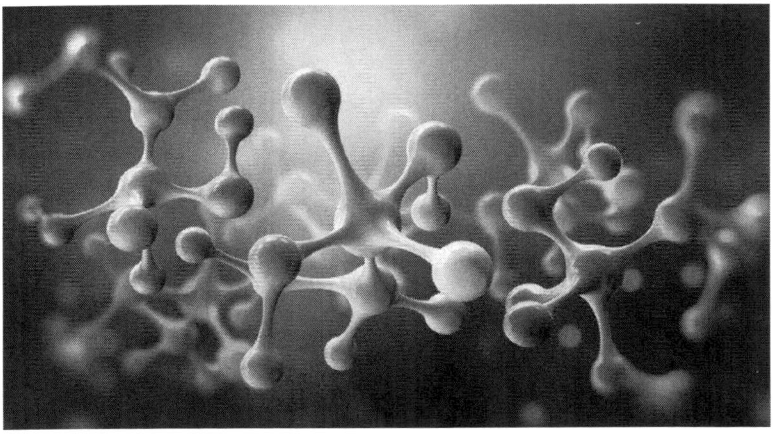

Fig 2. Nano size molecule, cell, crystal lattice.

Image Credit: Siarhei/ Shutterstock

The synthesis and characterization of nanomaterials is a critical component of nanochemistry. One of the challenges in synthesizing nanomaterials is controlling the size, shape, and composition of the particles. These factors can greatly affect the properties and functionalities of the materials. Various synthesis methods have been developed, including chemical vapor deposition, sol-gel synthesis, and electrodeposition, among others. Characterization of nanomaterials is also a critical step in understanding their properties and behavior. Techniques such as transmission electron microscopy (TEM), X-ray diffraction (XRD), and atomic force microscopy (AFM) are commonly used to characterize nanomaterials.

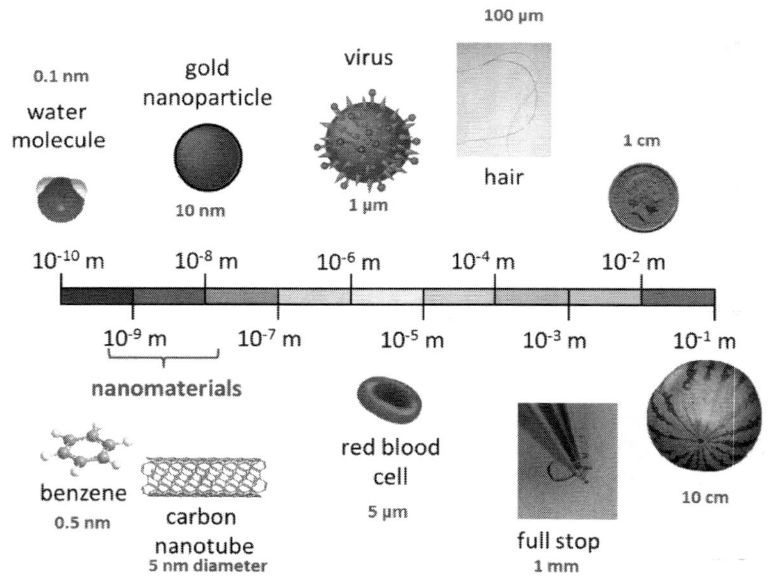

Fig 3. Nanosize metre

Another important aspect of nanochemistry is the study of the properties and behavior of nanomaterials. Due to their small size, nanomaterials exhibit unique properties that are different from bulk materials. For example, the optical properties of nanoparticles are highly dependent on their size and composition, allowing for the precise tuning of their properties. The electronic properties of nanomaterials are also highly dependent on their size and composition, and can exhibit quantum confinement effects. This leads to a wide range of potential applications in electronic devices, such as transistors and sensors. In addition, the high surface area-to-volume ratio of nanomaterials makes them highly reactive, leading to potential applications in catalysis and chemical sensing. The mechanical properties of nanomaterials are also important, as they can exhibit unique mechanical behavior due to their small size. For example, carbon nanotubes have been shown to be among the strongest materials known, with high tensile strength and stiffness.

In summary, nanochemistry is a rapidly growing field of science with diverse applications in electronics, energy, medicine, and catalysis, among others. The unique properties of nanomaterials make them attractive for a wide range of applications, and the synthesis, characterization, and manipulation of these materials is a critical aspect of nanochemistry. As the field continues to grow, new and exciting applications of nanomaterials are likely to emerge, making nanochemistry an exciting and dynamic field of research.

1.2 Definition and Scope of Nanochemistry

Nanochemistry is a field of chemistry that involves the study of chemical interactions and phenomena at the nanoscale level. The term "nano" refers to the size range of particles and materials that are typically between 1 and 100 nanometers in diameter. The field of nanochemistry encompasses the synthesis, characterization, and manipulation of nanoscale materials such as nanoparticles, nanotubes, and nanowires. This multidisciplinary field draws on chemistry, physics, materials science, and engineering to design and create materials with unique properties and functionalities.

The unique properties of nanomaterials arise due to their small size and large surface area-to-volume ratio. For example, the high surface area of nanoparticles allows for increased reactivity, and the quantum confinement of electrons in nanoscale materials leads to unique optical and electronic properties. These properties make nanomaterials attractive for use in a wide range of applications, including electronics, energy, medicine, and catalysis.

The scope of nanochemistry is broad and encompasses a wide range of research areas. One important area of research in nanochemistry is the synthesis of nanomaterials. The synthesis of nanomaterials involves the preparation of materials at the nanoscale level with controlled size, shape, and composition. Various synthesis methods have been developed, including chemical vapor deposition, sol-gel synthesis, and electrodeposition, among others. The control of these parameters is critical to achieving the desired properties and functionalities of the nanomaterials.

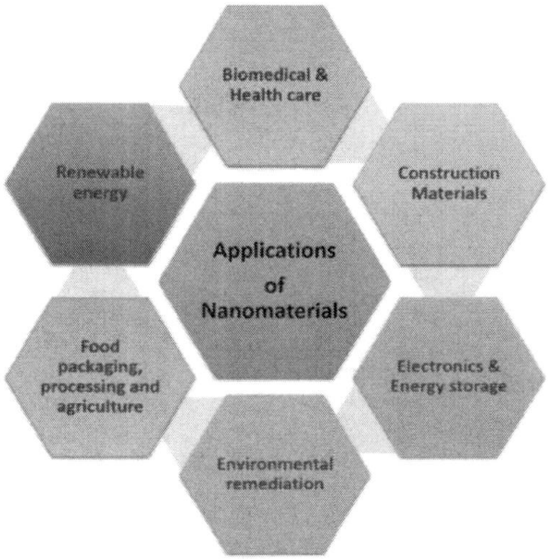

Fig 4. Applications of nanomaterials

Another important aspect of nanochemistry is the characterization of nanomaterials. The characterization of nanomaterials is a critical step in understanding their properties and behavior. Techniques such as transmission electron microscopy (TEM), X-ray diffraction (XRD), and atomic force microscopy (AFM) are commonly used to characterize nanomaterials. These techniques provide information on the size, shape, composition, and crystal structure of nanomaterials.

The study of the properties and behavior of nanomaterials is another important aspect of nanochemistry. Due to their small size, nanomaterials exhibit unique properties that are different from bulk materials. The optical properties of nanoparticles are highly dependent on their size and composition, allowing for the precise tuning of their properties. The electronic properties of nanomaterials are also highly dependent on their size and composition and can exhibit quantum confinement effects. This leads to a wide range of potential applications in electronic devices, such as transistors and sensors. The high surface area-to-volume ratio of nanomaterials makes them highly reactive, leading to potential applications in catalysis and chemical sensing. The mechanical properties of nanomaterials are also important, as they can exhibit unique mechanical behavior due to their small size.

One of the most important applications of nanomaterials is in the field of nanoelectronics. Nanoelectronics involves the design and fabrication of electronic devices and circuits at the nanoscale level. This is achieved using nanomaterials such as nanowires, nanotubes, and graphene, which have unique electronic properties due to their size and composition. These materials have been used to develop faster and more efficient computer processors and memory devices, and they hold great promise for the development of future electronic devices.

Another important application of nanomaterials is in the field of energy. Nanomaterials have been developed for use in solar cells, which convert sunlight into electricity, and for use in batteries and fuel cells, which store and release energy. For example, nanoparticles can be used as catalysts to speed up chemical reactions in fuel cells, making them more efficient.

Nanomaterials also have great potential in the field of medicine. Nanoparticles have been developed that can be used for targeted drug delivery, allowing for the precise delivery of drugs to specific cells or

1.3 Brief History of Nanochemistry

Nanochemistry is a relatively new field of research that emerged in the late 20th century, but the study of nanoscale materials and phenomena has a long history that dates back to ancient times.

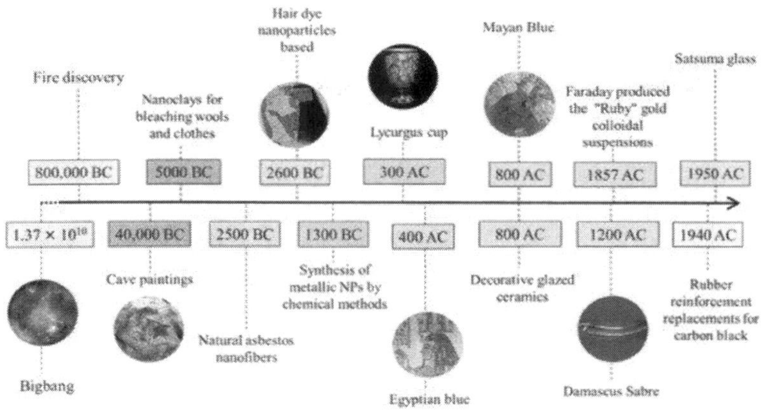

Fig 5. Historical timeline of nanochemistry

Here is a brief overview of the history of nanochemistry:

Ancient times: In ancient times, artisans and craftsmen were able to produce materials with nanoscale features through processes such as grinding, polishing, and glazing. For example, the Romans were known for their use of nanoscale particles of gold and silver in their glassware, which gave the glass a distinctive color.

Fig 6. Hourglass and book

17th century: In the 17th century, scientists began to study the properties of materials at the nanoscale. One notable example is Robert Boyle, who observed that the color of gold changes as its particle size is reduced.

- 19th century: In the 19th century, scientists such as Michael Faraday and James Clerk Maxwell made important contributions to the understanding of the behavior of materials at the nanoscale. Faraday's work on colloidal particles laid the foundation for the study of nanoparticles, while Maxwell's equations provided a framework for understanding the behavior of electromagnetic radiation, which is important for the study of optics and materials science.
- 1959: In 1959, physicist Richard Feynman gave a lecture titled "There's Plenty of Room at the Bottom," in which he discussed the possibility of manipulating individual atoms and molecules to create new materials and devices.
- 1981: In 1981, the term "nanotechnology" was coined by scientist K. Eric Drexler in his book "Engines of Creation: The Coming Era of Nanotechnology."
- 1985: In 1985, the first fullerene molecule was discovered by Harold Kroto, Robert Curl, and Richard Smalley. Fullerenes are a type of nanomaterial that consist of carbon atoms arranged in a hollow sphere or tube.
- 1991: In 1991, the first carbon nanotubes were discovered by Sumio Iijima. Carbon nanotubes are a type of nanomaterial that consist of carbon atoms arranged in a cylindrical structure.
- 1998: In 1998, researchers at IBM were able to use a scanning tunneling microscope to manipulate individual atoms, marking a major milestone in the development of nanotechnology.
- 2000s: In the 2000s, the field of nanochemistry began to emerge as a distinct area of research, with a focus on the synthesis, characterization, and applications of nanomaterials.

Overall, the history of nanochemistry is closely tied to the development of new technologies and scientific discoveries, and it is likely that the field will continue to evolve and expand in the coming years.

CHALLENGES

- What is the definition of a nanomaterial, and how is it different from bulk materials?
- What are some unique properties of nanomaterials that make them attractive for various applications?
- How did the study of nanomaterials evolve, and what are some key milestones in the history of this field?
- What is the role of nanomaterials in the development of nanotechnology, and what are some current and future applications of this technology?
- How are nanomaterials synthesized, and what are some common methods for producing them?
- What are some potential risks associated with the use and production of nanomaterials, and how can they be mitigated?
- What is the regulatory landscape for nanomaterials, and how are they currently classified and monitored?
- How do the properties of nanomaterials vary depending on their composition, structure, and size?
- What are some challenges and limitations associated with the synthesis and characterization of nanomaterials?
- How do the properties of nanomaterials change in different environments, such as in the presence of biological or chemical agents?

- What is the role of computational modeling in the study of nanomaterials, and how is it used to predict and optimize their properties?
- How do nanomaterials interact with biological systems, and what are some potential biomedical applications of this technology?
- What is the current state of commercialization of nanomaterials, and what are some industries that are leveraging this technology?
- How do societal factors influence the development and adoption of nanomaterials, and what are some ethical considerations related to this technology?
- What are some future directions for the study and application of nanomaterials, and how will this field evolve in the coming years?

CHAPTER TWO

2.1 Size and Shape of Nanoparticles

Nanoparticles are particles that range in size from 1 to 100 nanometers (nm) in diameter. At this size, the properties of materials can change significantly from their bulk form due to the increase in surface area-to-volume ratio, leading to unique physical and chemical properties. The size and shape of nanoparticles play a critical role in determining their properties and applications, and the synthesis of nanoparticles with specific size and shape is an important area of research in nanochemistry.

Size of nanoparticles

The size of nanoparticles has a significant impact on their properties, with smaller nanoparticles exhibiting different behavior than larger ones. This is due to the fact that the surface area-to-volume ratio of nanoparticles increases as their size decreases, leading to a higher number of surface atoms and greater reactivity.

One of the most significant changes in properties that occurs with decreasing nanoparticle size is a shift in the energy band structure of the material. In bulk materials, the energy bands are continuous, but in nanoparticles, the energy levels are quantized due to the confinement of the electrons in the small space of the nanoparticle. This leads to a shift in the electronic and optical properties of the material, with smaller nanoparticles exhibiting higher energy transitions and absorption/emission peaks.

Nanoparticles also exhibit size-dependent changes in their melting point, surface tension, and reactivity, which can have important implications for their use in applications such as catalysis and drug delivery.

Shape of nanoparticles

In addition to size, the shape of nanoparticles can also have a significant impact on their properties. Different shapes of nanoparticles can have different arrangements of atoms on their surface, leading to variations in surface energy, reactivity, and electronic properties.

One of the most well-studied shapes of nanoparticles is the sphere, which is commonly used in research as a model nanoparticle. Spherical nanoparticles have a high degree of symmetry, which leads to uniform surface properties and a high degree of stability. However, other shapes of nanoparticles, such as rods, cubes, and tetrapods, can exhibit enhanced properties due to their anisotropic structure.

For example, rod-shaped nanoparticles have been shown to exhibit enhanced plasmonic properties compared to spherical nanoparticles due to their elongated shape. Plasmonic nanoparticles have applications in sensing, imaging, and optical communication due to their ability to manipulate light.

Similarly, cubic nanoparticles have been shown to exhibit enhanced catalytic activity compared to spherical nanoparticles due to the exposed facets on their surfaces. The exposed facets can have different atomic arrangements and surface energies, leading to variations in reactivity and selectivity in catalytic reactions.

Particle Form	Shapes	Shape Factor	Particle Form	Shapes	Shape Factor
Sphere		3	Tetrahedron	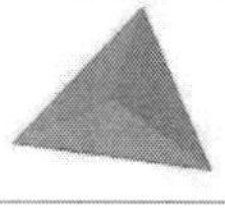	4.06
Bricks		3.7	Hexahedron		3.72
Cylinder		4.9	Octahedron	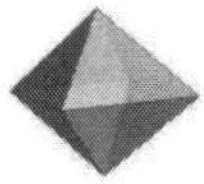	1.18
Platelet		5.7	Blade		8.9

Fig 7. Shapes of nanomaterials

Synthesis of nanoparticles with specific size and shape:

The ability to synthesize nanoparticles with specific size and shape is critical for controlling their properties and applications. There are a number of methods for synthesizing nanoparticles, including physical methods such as laser ablation and chemical methods such as precipitation and sol-gel synthesis.

In many cases, the synthesis of nanoparticles with specific size and shape is achieved through the use of templating agents.

Templating agents are molecules or templates that direct the formation of nanoparticles with specific sizes and shapes. For example, surfactants can be used to control the size and shape of gold nanoparticles during synthesis, while dendrimers can be used to control the size and shape of nanoparticles in solution.

Fig 8. Dendrimers

Another important area of research in the synthesis of nanoparticles is the use of bottom-up approaches, in which individual atoms or molecules are assembled into nanoparticles using chemical reactions. This approach has the potential to produce nanoparticles with a high degree of control over their size and shape, but it is still a challenging area of research due to the complexity of the chemical reactions involved.

- Nano-sized particles: 1-100 nanometers in diameter
- Nanocrystals or quantum dots: 1-20 nanometers in size
- Nanowires: typically less than 100 nanometers in diameter and up to several micrometers in length
- Nanotubes: hollow structures with diameters ranging from 1-100 nanometers and lengths up to several centimeters
- Nanoporous materials: materials with pores or cavities on the nanometer scale, typically with sizes ranging from 1-100 nanometers
- Nanocomposites: materials consisting of a combination of nanoparticles and a bulk material, typically with particle sizes ranging from 1-100 nanometers.

2.2 Surface Area and Surface Energy of Nanoparticles

The surface area and surface energy of nanomaterials are critical parameters that can significantly influence their properties and behavior. At the nanoscale, the surface area-to-volume ratio of materials increases significantly, leading to a higher number of surface atoms and greater surface energy. This high surface energy can lead to unique physical and chemical properties, such as enhanced reactivity, catalytic activity, and adsorption capacity.

Surface Area

The surface area of a nanomaterial refers to the total area of its exposed surfaces, which increases significantly as the particle size decreases. For example, a 1 mm cube has a total surface area of 6 mm^2, while a 1 nm cube has a total surface area of 6 x 10^5 m^2. This increase in surface area can lead to enhanced reactivity, as a larger number of atoms are exposed to the surrounding environment.

The high surface area of nanomaterials can also lead to increased surface reactivity and sensitivity to environmental factors such as temperature, pressure, and humidity. This can be beneficial in applications such as catalysis and sensing, where the reactivity and sensitivity of the material are critical.

Surface Energy

The surface energy of a nanomaterial refers to the energy required to create a new surface, and is related to the cohesive energy of the material. At the nanoscale, the surface energy can become a dominant factor in determining the properties and behavior of materials.

The high surface energy of nanomaterials can lead to enhanced adsorption capacity, as molecules can be attracted to the surface of the material due to the high surface energy. This property can be utilized in applications such as environmental remediation and drug delivery, where the adsorption of contaminants or therapeutic molecules is important.

The surface energy of nanomaterials can also influence their stability, as materials with high surface energy are more likely to agglomerate or undergo surface reactions. This can be addressed through the use of surface coatings or modifications, which can reduce the surface energy of the material and improve its stability.

Overall, the surface area and surface energy of nanomaterials are critical parameters that must be carefully considered in the design and synthesis of nanomaterials for specific applications.

The unique properties and behavior of nanomaterials at the nanoscale make them a promising area of research for a wide range of applications in fields such as electronics, energy, medicine, and materials science.

2.3 Optical and Electronic Properties of Nanoparticles

Optical properties

The optical properties of nanomaterials refer to their interaction with light and include phenomena such as absorption, reflection, and scattering. The size and shape of the nanomaterials strongly affect their optical properties due to the quantum confinement effects. These effects can cause the energy levels of the electrons to become quantized and result in a shift in the absorption or emission spectra of the material.

Nanomaterials can exhibit size-dependent optical properties, such as tunable absorption and emission spectra. For example, quantum dots are a type of nanomaterial that exhibit size-dependent emission spectra due to their quantum confinement effects. They emit light of a specific color when excited by a particular wavelength of light, which makes them useful in applications such as biological imaging and solid-state lighting.

Here are some of the optical properties of nanomaterials:

a. Absorption: Nanomaterials can absorb light of specific wavelengths depending on their size and shape. The absorption spectra of nanomaterials can shift due to quantum confinement effects, leading to tunable absorption properties.
b. Emission: Nanomaterials can emit light of specific colors when excited by a particular wavelength of light. This

property is often used in biological imaging and solid-state lighting.

c. Plasmon resonance: Metal nanoparticles can exhibit localized surface plasmon resonance, which is a collective oscillation of the conduction electrons in response to an incident electromagnetic field. This property gives rise to intense absorption and scattering of light, making metal nanoparticles useful in applications such as sensing and imaging.
d. Surface-enhanced Raman scattering (SERS): Nanomaterials can enhance the Raman scattering signal of molecules adsorbed on their surface due to plasmon resonance effects. This property is used in sensing and detection applications.
e. Photoluminescence: Some nanomaterials, such as quantum dots and metal chalcogenides, exhibit photoluminescence, which is the emission of light by a material after absorbing light. This property is used in applications such as lighting and display technologies.
f. Nonlinear optical properties: Nanomaterials can exhibit nonlinear optical properties, which means that the response of the material to an incident electromagnetic field is not proportional to the intensity of the field. This property is useful in applications such as nonlinear optics and laser technology.
g. Chiral properties: Chiral nanomaterials, such as chiral nanoparticles and nanotubes, exhibit circular dichroism, which is a difference in absorption of left- and right-handed circularly polarized light. This property is used in sensing and chiral separation applications.

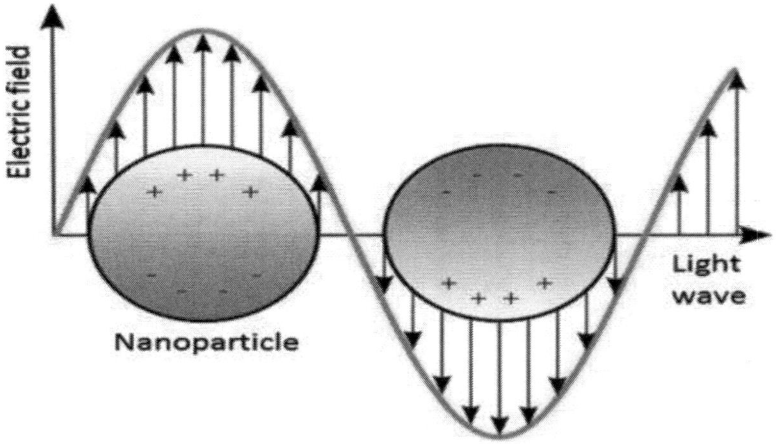

Fig 9. Optical properties of nanomaterials

Overall, the optical properties of nanomaterials are critical in determining their performance in various applications, and their unique properties at the nanoscale make them attractive candidates for developing new technologies in fields such as optics, sensing, and imaging.

Electronic Properties

The electronic properties of nanomaterials refer to their ability to conduct or insulate electric current. These properties are strongly influenced by the size and shape of the nanomaterial, which can cause quantum confinement effects and change the band structure of the material.

Here is a list of electronic properties of nanomaterials:

a. Quantum confinement: The electronic properties of nanomaterials are strongly influenced by quantum confinement, which arises due to the confinement of electrons within a small volume of space.

b. Size-dependent bandgap: The bandgap of nanomaterials is size-dependent and can be tuned by controlling the size and shape of the nanoparticles. Small nanoparticles have a larger bandgap than larger ones.
c. Plasmon resonance: Nanoparticles exhibit plasmon resonance, which arises due to the collective oscillation of electrons on the surface of the nanoparticles. This property is useful in applications such as sensing and imaging.
d. Surface states: The electronic states at the surface of nanoparticles are different from those in the bulk material, and these surface states can have a significant impact on the electronic properties of the material.
e. Quantum tunneling: Quantum tunneling is the process by which electrons can tunnel through a potential barrier. This phenomenon is more pronounced in nanomaterials due to the reduced dimensions.
f. Coulomb blockade: Coulomb blockade is a quantum phenomenon that arises in small electronic systems, where the flow of electrons is blocked by Coulomb repulsion.
g. Charge transfer: The electronic properties of nanomaterials can be tuned by charge transfer between the nanoparticles and other materials. This property is useful in applications such as solar cells and photocatalysis.
h. Surface plasmon resonance: Surface plasmon resonance is a phenomenon that arises due to the collective oscillation of electrons on the surface of metal nanoparticles. This property is useful in applications such as biosensors and catalysis.
i. Carrier mobility: The mobility of charge carriers in nanomaterials is higher than that in bulk materials, due to the reduced dimensions and increased surface area.
j. Energy transfer: The electronic properties of nanomaterials can be tuned by energy transfer between the nanoparticles and other materials. This property is useful in applications such as light harvesting and energy storage.

Nanomaterials can exhibit unique electronic properties, such as high conductivity or insulating behavior, depending on their size and composition. For example, carbon nanotubes are a type of nanomaterial that exhibit high electrical conductivity due to their one-dimensional structure and high surface area. They are used in applications such as electronic devices and energy storage.

Nanomaterials can also exhibit unique properties such as tunneling conductance, spin effects, and non-linear optical behavior, which can be utilized in various applications such as quantum computing, electronics, and photonics.

Overall, the optical and electronic properties of nanomaterials are critical factors in determining their performance in various applications. The unique properties of nanomaterials at the nanoscale make them attractive candidates for developing new technologies in fields such as electronics, photonics, and energy. However, the study and understanding of these properties require specialized techniques and equipment, which continue to be developed by researchers in the field of nanomaterials.

2.4 Thermal and Mechanical Properties of Nanoparticles

Nanomaterials exhibit unique thermal and mechanical properties that differ significantly from their bulk counterparts due to their small size and high surface area-to-volume ratio. Here are some key points about the thermal and mechanical properties of nanomaterials:

Thermal properties

Nanomaterials have higher melting points and thermal stability than bulk materials due to the reduced number of surface atoms and enhanced bonding at the surface. They also exhibit enhanced thermal conductivity, which can be useful for applications in electronics, energy storage, and heat transfer.

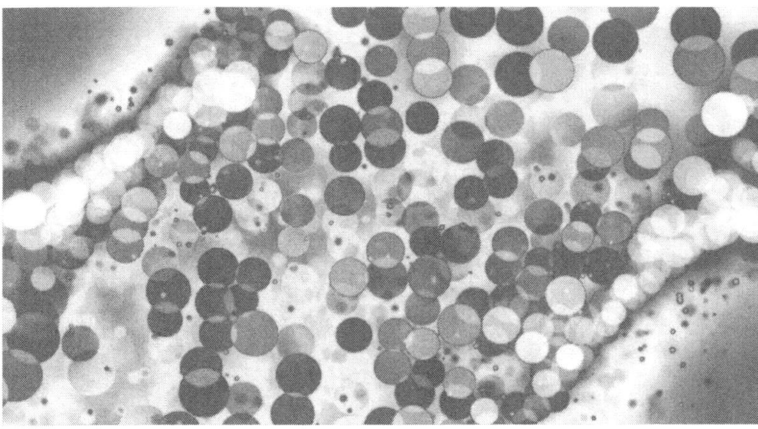

Fig 10. Heat transfer properties of nanoparticles

Here are some thermal properties of nanomaterials:

a. Melting point: The melting point of nanomaterials can be different from bulk materials due to the change in surface area to volume ratio.
b. Thermal conductivity: The thermal conductivity of nanomaterials can be higher than bulk materials due to the increase in surface area.
c. Specific heat capacity: The specific heat capacity of nanomaterials can be different from bulk materials due to the changes in size and shape.
d. Phase transitions: Nanomaterials can undergo phase transitions at lower temperatures than bulk materials due to their small size.

e. Thermal stability: The thermal stability of nanomaterials can be different from bulk materials due to the changes in surface energy and crystal structure.
f. Thermoelectric properties: Nanomaterials can exhibit enhanced thermoelectric properties due to the changes in electronic structure and phonon transport.
g. Thermal expansion: The thermal expansion of nanomaterials can be different from bulk materials due to the changes in crystal structure and defects.
h. Thermal shock resistance: Nanomaterials can exhibit improved thermal shock resistance due to the changes in surface energy and crystal structure.
i. Heat capacity enhancement: The heat capacity of nanomaterials can be enhanced due to the changes in surface area and electronic structure.
j. High-temperature stability: Nanomaterials can exhibit high-temperature stability due to the changes in crystal structure and surface energy.

Mechanical properties

Nanomaterials have higher strength, hardness, and elastic modulus than bulk materials due to the presence of defects and grain boundaries that hinder dislocation motion and improve strength. They also exhibit enhanced ductility, toughness, and fatigue resistance, which can be useful for applications in structural materials, sensors, and biomedical implants.

Here are some examples of mechanical properties of nanomaterials:

a. Hardness: Nanomaterials have higher hardness than their bulk counterparts due to their high surface area-to-volume ratio, which causes a higher concentration of dislocations and defects that strengthen the material.

b. Elasticity: Nanomaterials have unique elastic properties due to their small size and high surface area-to-volume ratio. They exhibit higher stiffness and elastic modulus compared to their bulk counterparts.
c. Ductility: Some nanomaterials such as nanowires and nanotubes exhibit high ductility due to their elongated shape and flexibility.
d. Toughness: Nanomaterials can exhibit higher toughness due to their unique deformation behavior at the nanoscale. The high density of dislocations and defects can result in energy dissipation during deformation, which can increase the toughness of the material.
e. Fatigue resistance: Nanomaterials can exhibit improved fatigue resistance due to their high surface area-to-volume ratio, which reduces the likelihood of crack initiation and propagation.
f. Creep resistance: Some nanomaterials such as nanocomposites can exhibit improved creep resistance due to their unique microstructure and interface properties.
g. Wear resistance: Nanomaterials can exhibit improved wear resistance due to their high hardness and unique deformation behavior at the nanoscale. They can also exhibit self-lubricating behavior due to their high surface area-to-volume ratio and surface chemistry.
h. Fracture toughness: Nanomaterials can exhibit higher fracture toughness compared to their bulk counterparts due to their unique deformation behavior and the presence of defects at the nanoscale. The small size of the defects can hinder crack propagation and increase the material's fracture toughness.

CHALLENGES

- What are some examples of optical properties that can be enhanced in nanomaterials compared to bulk materials?
- How can the size and shape of nanoparticles affect their optical properties?
- What is plasmon resonance in nanoparticles, and how does it relate to their optical properties?
- How do quantum confinement effects influence the absorption and emission spectra of nanomaterials?
- What are some potential applications of nanomaterials with unique optical properties, such as quantum dots or nanophosphors?
- What are some examples of electrical properties that can be enhanced in nanomaterials compared to bulk materials?
- How does the size and shape of nanoparticles affect their electrical conductivity and resistance?
- What is the difference between intrinsic and extrinsic doping of nanomaterials, and how does it affect their electrical properties?
- How do quantum confinement effects influence the electronic structure and transport properties of nanomaterials?
- What are some potential applications of nanomaterials with unique electrical properties, such as carbon nanotubes, graphene, or quantum wires?

- How does the thermal conductivity of nanomaterials differ from that of bulk materials, and what are some factors that influence it?
- What is the effect of size and shape on the thermal properties of nanoparticles and nanowires?
- How do phonon confinement effects influence the thermal conductivity of nanomaterials, and what are some methods to control it?
- What is the difference between electronic and lattice thermal conductivity in nanomaterials, and how do they interact?
- What are some potential applications of nanomaterials with unique thermal properties, such as thermoelectric materials, thermal barrier coatings, or phase change materials?
- How do the mechanical properties of nanomaterials differ from those of bulk materials, and what are some factors that influence them?
- What is the effect of size and shape on the mechanical properties of nanoparticles and nanowires?
- How do defects and grain boundaries influence the mechanical behavior of nanomaterials?
- What is the difference between brittle and ductile behavior in nanomaterials, and how does it relate to their structure and properties?
- What are some potential applications of nanomaterials with unique mechanical properties, such as high-strength materials, self-healing materials, or shape memory alloys?

CHAPTER THREE

3.1 Synthesis of Nanoparticles

The synthesis of nanomaterials can be achieved through various methods, including physical, chemical, and biological methods. Physical methods involve the use of physical forces, such as milling and lithography, to reduce the size of materials to the nanoscale. Chemical methods, on the other hand, involve the use of chemical reactions to create nanomaterials, such as the reduction of metal ions to form metal nanoparticles. Biological methods involve the use of living organisms or their byproducts to synthesize nanoparticles.

Each method of nanomaterial synthesis has its advantages and disadvantages, depending on the type of material being synthesized, the desired properties, and the intended applications. For example, physical methods are suitable for synthesizing inorganic nanomaterials, such as metals and metal oxides, but may not be suitable for creating complex structures or functionalizing surfaces. Chemical methods are suitable for creating a wide range of nanomaterials with different properties, but may require harsh chemical conditions or produce toxic byproducts. Biological methods are sustainable and environmentally friendly, but may require longer reaction times and may not be suitable for creating certain types of nanomaterials.

In recent years, the synthesis of nanomaterials has become an active area of research, as scientists and engineers strive to create materials with new or improved properties for various applications. The synthesis of nanomaterials is a complex and interdisciplinary field that involves chemistry, physics, biology, and engineering. Researchers continue to explore new methods and techniques to create nanomaterials with precisely controlled properties, while also addressing the challenges of scalability, reproducibility, and safety.

3.2 Methods of Nanomaterial Synthesis

Physical Methods of Synthesis

There are several physical methods of synthesizing nanomaterials, including:

a. Vapor phase methods: In vapor phase methods, the precursor material is evaporated, and the resulting vapor is condensed onto a substrate or collected in a chamber. Vapor phase methods include techniques such as physical vapor deposition (PVD) and chemical vapor deposition (CVD).
b. Laser ablation: Laser ablation involves using a high-powered laser to vaporize a target material, which then condenses to form nanoparticles. This technique can be used to synthesize a wide range of materials, including metals, oxides, and semiconductors.
c. Plasma synthesis: Plasma synthesis involves using a plasma to generate reactive species, which can then react to form nanoparticles. This technique can be used to synthesize a wide range of materials, including metals, oxides, and carbon-based materials.

d. High-energy ball milling: High-energy ball milling involves using a ball mill to mechanically grind the precursor material, resulting in the formation of nanoparticles. This technique can be used to synthesize a wide range of materials, including metals, oxides, and semiconductors.
e. Electrochemical synthesis: Electrochemical synthesis involves using an electric current to drive the formation of nanoparticles. This technique can be used to synthesize a wide range of materials, including metals, oxides, and semiconductors.
f. Sputtering: Sputtering involves bombarding a target material with high-energy ions, causing atoms to be ejected and deposited onto a substrate to form nanoparticles. This technique can be used to synthesize a wide range of materials, including metals, oxides, and semiconductors.
g. Each of these physical methods has its advantages and limitations, depending on the specific application and desired properties of the nanomaterial. For example, vapor phase methods are useful for synthesizing materials with high purity and control over size and morphology, while high-energy ball milling is useful for producing nanoparticles in large quantities. Overall, physical methods of synthesis are important for producing nanomaterials with specific properties and functionalities for a wide range of applications, including electronics, energy, and biomedical engineering.

Chemical Methods of Synthesis

Chemical methods are one of the most commonly used methods for synthesizing nanomaterials, which involve chemical reactions that lead to the formation of nanoparticles. Some of the chemical methods used for the synthesis of nanomaterials are as follows:

a. Sol-gel method: In the sol-gel method, a sol, which is a colloidal suspension of nanoparticles, is prepared by hydrolyzing metal alkoxides in a solvent. The sol is then converted into a gel by aging and drying, followed by calcination to produce nanoparticles. This method is widely used for synthesizing metal oxide nanoparticles.
b. Chemical precipitation: Chemical precipitation is a simple and cost-effective method for synthesizing metal and metal oxide nanoparticles. In this method, metal salts are dissolved in a solvent, followed by the addition of a reducing agent or a base to initiate the precipitation of nanoparticles. The size and morphology of the nanoparticles can be controlled by adjusting the reaction parameters.
c. Hydrothermal/solvothermal synthesis: Hydrothermal or solvothermal synthesis involves the use of high-pressure and high-temperature conditions to promote the growth of nanoparticles. In this method, a precursor solution is heated under pressure, leading to the nucleation and growth of nanoparticles. This method is widely used for synthesizing metal oxide nanoparticles.
d. Microemulsion: Microemulsion is a method for synthesizing nanoparticles using a microemulsion, which is a stable mixture of water, oil, and surfactant. In this method, a precursor solution is added to the microemulsion, followed by the addition of a reducing agent to initiate the formation of nanoparticles. This method is useful for synthesizing metal nanoparticles.
e. Electrochemical synthesis: Electrochemical synthesis involves the use of an electric current to drive the formation of nanoparticles. In this method, the precursor material is dissolved in an electrolyte solution, and an electric current is applied to initiate the formation of nanoparticles. This method is useful for synthesizing metal and metal oxide nanoparticles.
f. Spray pyrolysis: Spray pyrolysis is a method for synthesizing nanoparticles by spraying a precursor solution

onto a hot surface, causing it to evaporate and react to form nanoparticles. This method is useful for synthesizing metal and metal oxide nanoparticles.

Overall, chemical methods are widely used for the synthesis of nanomaterials due to their simplicity, cost-effectiveness, and scalability. However, the choice of the method depends on the desired properties of the nanomaterial, including size, shape, and composition.

Biological Methods of Synthesis

Biological methods of nanomaterial synthesis involve the use of living organisms or their byproducts to synthesize nanoparticles. These methods have gained increasing attention in recent years due to their sustainability, low cost, and biocompatibility. Some of the commonly used biological methods for synthesizing nanomaterials are as follows:

a. Microbial synthesis: Microbial synthesis involves the use of microorganisms, such as bacteria, fungi, and algae, to synthesize nanoparticles. In this method, microorganisms are grown in a nutrient medium containing metal ions, which are then reduced and stabilized on the microbial cell surface to form nanoparticles. This method is useful for synthesizing metal and metal oxide nanoparticles.
b. Plant-mediated synthesis: Plant-mediated synthesis involves the use of plant extracts or plant parts, such as leaves, stems, and roots, to synthesize nanoparticles. In this method, plant extracts containing phytochemicals, such as flavonoids and terpenoids, are mixed with metal ions, leading to the reduction and stabilization of nanoparticles. This method is useful for synthesizing metal and metal oxide nanoparticles.
c. Enzymatic synthesis: Enzymatic synthesis involves the use of enzymes to synthesize nanoparticles. In this method,

enzymes, such as peroxidase and alkaline phosphatase, are used as reducing agents to convert metal ions into nanoparticles. This method is useful for synthesizing metal and metal oxide nanoparticles.

d. Biomineralization: Biomineralization involves the use of biological macromolecules, such as proteins and peptides, to synthesize nanoparticles. In this method, the biological macromolecules act as templates for the nucleation and growth of nanoparticles, leading to the formation of highly uniform and stable nanoparticles. This method is useful for synthesizing metal and metal oxide nanoparticles.

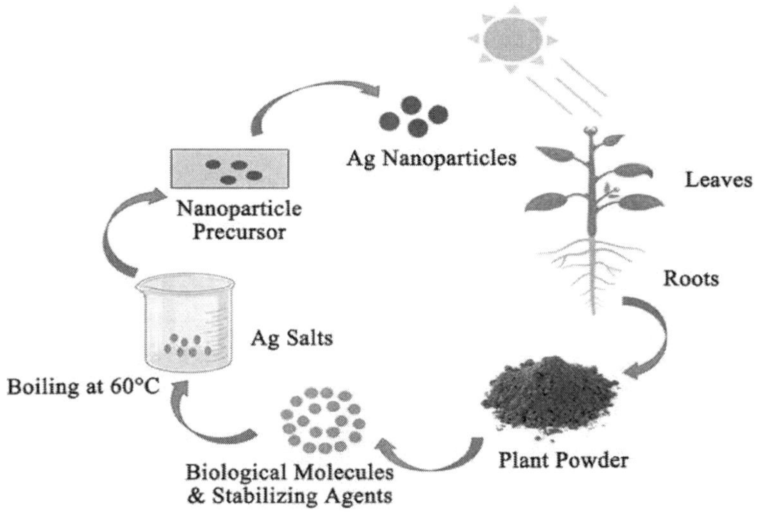

Fig 11. Biological synthesis of nanomaterials

Overall, biological methods of nanomaterial synthesis offer several advantages over traditional chemical methods, including eco-friendliness, biocompatibility, and the ability to produce highly uniform and stable nanoparticles. These methods have found applications in various fields, such as biomedicine, environmental remediation, and energy storage.

3.3 Approaches To Nanomaterial Synthesis

There are two approaches to nanomaterial synthesis: 1. Top-down synthesis 2. Bottom-up synthesis.

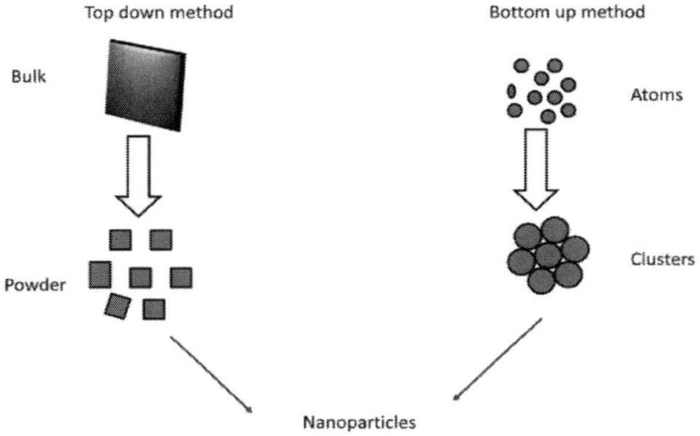

Fig 12. Approaches to nanomaterials synthesis

Top-down approach

Top-down synthesis is a physical method for synthesizing nanomaterials that involves the mechanical fragmentation or etching of bulk materials to create smaller structures with dimensions in the nanometer range. This method involves the reduction of the size of materials through mechanical forces, such as milling or lithography, to create nanoscale structures.

In top-down synthesis, the starting material is usually a bulk material that is processed using mechanical or chemical means to create the desired nanoscale structures. The mechanical processes used in top-down synthesis include milling, grinding, cutting, and lithography. These processes can be used to create nanostructures such as nanoparticles, nanowires, and nanosheets.

Top-down synthesis is advantageous in that it allows for the creation of nanomaterials with well-defined shapes and sizes. The size and shape of the resulting nanomaterials can be controlled by adjusting the processing conditions, such as the milling time and speed, or the lithography pattern. Additionally, top-down synthesis is a scalable method that can be used to produce large quantities of nanomaterials.

However, top-down synthesis has its limitations. One of the main challenges of this method is the difficulty in creating complex nanostructures with precise control over their composition and morphology. The mechanical forces used in top-down synthesis can also result in defects and impurities, which can affect the properties of the resulting nanomaterials.

Overall, top-down synthesis is an important method for the synthesis of nanomaterials, particularly for inorganic materials such as metals and metal oxides. However, it is important to carefully consider the limitations and advantages of this method when choosing a synthesis approach for a particular applicationCharacterization of Nanoparticles.

Bottom-up approach

Bottom-up synthesis is a chemical method for synthesizing nanomaterials that involves the assembly of smaller building blocks into larger structures with dimensions in the nanometer range. This method involves the use of chemical reactions to create nanoscale structures from molecular or atomic precursors.

In bottom-up synthesis, the starting material is usually a molecular or atomic precursor that undergoes a chemical reaction to form nanoscale structures. The chemical processes used in bottom-up synthesis include chemical vapor deposition (CVD), sol-gel synthesis, and self-assembly.

Bottom-up synthesis has the advantage of allowing for precise control over the composition, structure, and morphology of the resulting nanomaterials. The chemical reactions used in bottom-up synthesis can be carefully tuned to produce nanomaterials with specific properties and functions. Additionally, bottom-up synthesis can be used to create complex nanostructures with high precision.

However, bottom-up synthesis also has its limitations. One of the main challenges of this method is the difficulty in scaling up the synthesis process to produce large quantities of nanomaterials. Additionally, the synthesis conditions must be carefully controlled to ensure reproducibility and prevent the formation of impurities and defects.

Overall, bottom-up synthesis is an important method for the synthesis of nanomaterials, particularly for organic and hybrid materials. However, it is important to carefully consider the limitations and advantages of this method when choosing a synthesis approach for a particular application.

CHALLENGES

- What are some common types of nanomaterials, and how are they synthesized?
- How do the properties of nanomaterials vary depending on their synthesis method?
- What are some advantages and disadvantages of top-down and bottom-up approaches to nanomaterials synthesis?
- What are some common top-down methods for nanomaterials synthesis, and how do they work?
- What are some common bottom-up methods for nanomaterials synthesis, and how do they work?
- How do hydrothermal and solvothermal synthesis methods work, and what are some applications of these approaches?
- What is the role of chemical vapor deposition in the synthesis of nanomaterials, and how is it used to produce thin films and coatings?
- What is the difference between wet and dry synthesis methods for nanomaterials, and what are some examples of each?
- What is the role of templating in nanomaterials synthesis, and how is it used to control particle size and shape?
- How does the use of surfactants and stabilizers influence the synthesis and properties of nanomaterials?
- What are some green synthesis methods for nanomaterials, and how do they compare to conventional methods in terms of sustainability?

- What is the role of sonochemistry in nanomaterials synthesis, and how is it used to promote particle formation and control size and shape?
- How does the use of external stimuli, such as light or heat, influence the synthesis and properties of nanomaterials?
- What is the role of microwave-assisted synthesis in nanomaterials production, and how does it compare to other synthesis methods?
- What are some challenges associated with scaling up nanomaterials synthesis from laboratory to industrial scale?
- How do self-assembly methods work in the synthesis of nanomaterials, and what are some advantages of this approach?
- What is the role of seed-mediated growth in the synthesis of nanomaterials, and how is it used to control particle size and morphology?
- How does the use of solvents and reaction conditions influence the synthesis and properties of nanomaterials?
- What is the role of nanolithography in nanomaterials synthesis, and how is it used to create patterned surfaces and devices?
- What are some emerging approaches to nanomaterials synthesis, and how do they differ from conventional methods?

CHAPTER FOUR

4.1 Analysis of Nanoparticles

Imaging Techniques for Nanoparticles

Imaging techniques play a crucial role in the characterization and analysis of nanomaterials, allowing researchers to visualize and study the properties of materials at the nanoscale.

Fig 13. Transmission Electron Microscope

Here are some common imaging techniques used for nanomaterials:

a. Scanning Electron Microscopy (SEM): SEM is a surface imaging technique that uses a beam of electrons to scan the surface of a sample, producing a 3D image of the surface morphology. SEM is widely used for the characterization of nanomaterials and nanostructures, providing information on their size, shape, and surface properties.
b. Transmission Electron Microscopy (TEM): TEM is a high-resolution imaging technique that uses a beam of electrons to transmit through a thin sample, producing an image of the material's internal structure. TEM is used to study the morphology, crystal structure, defects, and composition of nanomaterials.
c. Atomic Force Microscopy (AFM): AFM is a high-resolution imaging technique that uses a sharp tip to scan the surface of a sample, producing a 3D image of the surface morphology. AFM is used to study the size, shape, and surface properties of nanomaterials, as well as their mechanical and electrical properties.
d. Scanning Probe Microscopy (SPM): SPM is a family of high-resolution imaging techniques that use a probe to scan the surface of a sample, producing an image of the surface morphology. SPM includes AFM, STM (Scanning Tunneling Microscopy), and SNOM (Scanning Near-field Optical Microscopy), and is widely used for the characterization and manipulation of nanomaterials and nanostructures.
e. Confocal Microscopy: Confocal microscopy is a technique that uses a laser light source to scan a sample, producing high-resolution images of the sample's 3D structure. Confocal microscopy is used to study the size, shape, and distribution of nanoparticles, as well as their interactions with biological systems.
f. Scanning Tunneling Microscopy (STM) is a powerful imaging technique that has made significant contributions

to the field of nanotechnology and nanoscience. It was invented by Gerd Binnig and Heinrich Rohrer in 1981 and was awarded the Nobel Prize in Physics in 1986. The technique involves scanning the surface of a material using a sharp metallic tip with a nanometer-sized apex that is placed in close proximity to the sample surface. A voltage is applied between the tip and the sample, and the resulting tunneling current is measured as the tip is scanned over the surface. STM has several advantages over other imaging techniques. It provides high-resolution images of surfaces with atomic-level resolution and can also measure local electronic properties, such as the density of states and the local work function of the material. STM can be used to image a wide range of materials, including metals, semiconductors, insulators, and biological samples, making it a versatile tool for materials characterization. STM has also been used in nanotechnology for the fabrication of nanostructures. The technique of scanning tunneling lithography (STL) involves using the STM tip to modify the surface of a material by removing or rearranging atoms. STL has been used to create a variety of nanostructures, including quantum dots, nanowires, and nanorings, with precise control over their size, shape, and orientation.

g. Fluorescence microscopy is a technique widely used in nanotechnology to study the optical properties of nanoparticles and other nanostructures. This method uses fluorescent probes or dyes to label and detect specific molecules or structures in a sample, allowing for the visualization and quantification of biological, chemical, and physical processes at the nanoscale.

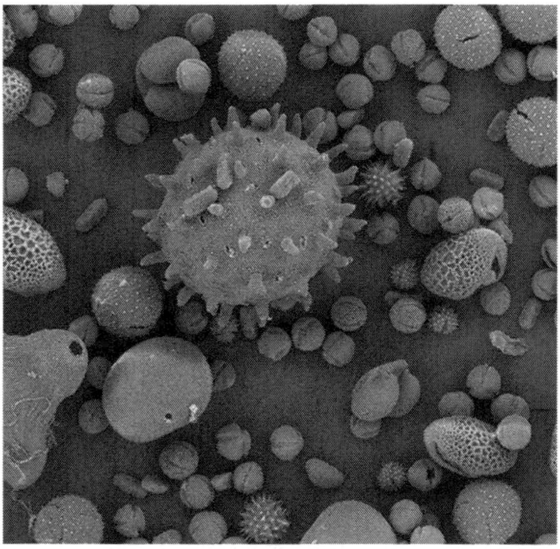

Fig 14. SEM analysis of nanoparticles

Spectroscopy Techniques for Nanoparticles

Spectroscopy techniques are widely used in nanomaterials research to investigate the optical and electronic properties of nanoparticles, as well as their chemical composition and structure.

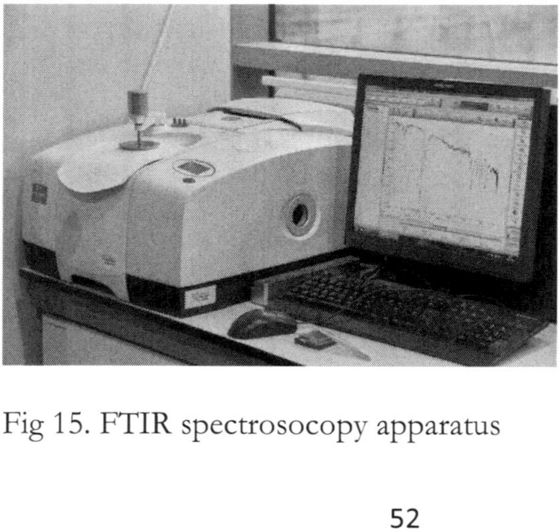

Fig 15. FTIR spectrosocopy apparatus

Here are some common spectroscopy techniques used in nanomaterials research:

a. UV-Vis spectroscopy: UV-Vis spectroscopy is a widely used technique to study the optical properties of nanomaterials by measuring their absorption and reflection of light in the ultraviolet and visible regions of the electromagnetic spectrum. It can provide information on the size, shape, composition, and surface chemistry of nanoparticles, as well as their plasmonic properties and surface-enhanced Raman scattering (SERS) activity. UV-Vis spectroscopy is particularly useful for the characterization of metal nanoparticles, quantum dots, and other optically active nanomaterials in various applications such as sensing, imaging, and catalysis
b. X-ray photoelectron spectroscopy (XPS): a surface-sensitive technique that measures the chemical composition and electronic state of materials, useful for characterizing nanoscale surfaces and interfaces.
c. Raman spectroscopy: a non-destructive technique that provides information on the vibrational modes of molecules in nanomaterials, providing information on structure, composition, and defects.
d. Fourier-transform infrared spectroscopy (FTIR): a technique that uses infrared light to measure the vibrational modes of molecules, useful for determining the chemical composition and functional groups in nanomaterials.
e. Energy-dispersive X-ray spectroscopy (EDS): a technique that detects X-rays emitted from nanomaterials excited by an electron beam, providing information on the elemental composition of the material.
f. X-ray diffraction (XRD): a technique that measures the diffraction of X-rays by crystal structures, providing information on crystallographic structure and phase purity of nanomaterials.

g. Dynamic light scattering (DLS): a technique also called Photon Correlation Spectroscopy measures the Brownian motion of nanoparticles in a solution, providing information on particle size, size distribution, and zeta potential.
h. Inductively coupled plasma mass spectrometry (ICP-MS): a technique that measures the elemental composition of nanomaterials in solution, useful for determining the amount and size distribution of nanoparticles in a sample.
i. Auger electron spectroscopy (AES): a surface-sensitive technique that measures the kinetic energy of electrons emitted from nanomaterials, providing information on the surface composition and chemistry of materials.
j. Laser-induced breakdown spectroscopy (LIBS): a technique that uses laser pulses to create a plasma on the surface of nanomaterials, providing information on elemental composition and concentration.
k. Time-of-flight secondary ion mass spectrometry (TOF-SIMS): a technique that provides information on the chemical composition and molecular structure of nanomaterials at the surface, useful for determining the chemical properties of nanoparticles.
l. X-ray absorption spectroscopy (XAS): a technique that measures the absorption of X-rays by nanomaterials, providing information on the chemical environment and electronic structure of materials.
m. Surface-enhanced Raman spectroscopy (SERS): a technique that enhances the Raman signal of molecules on the surface of nanomaterials, providing information on the chemical composition and molecular structure of materials at the surface.
n. Near-field scanning optical microscopy (NSOM): a technique that uses a nanoscale probe to scan the surface of nanomaterials, providing information on the optical and electronic properties of materials at the nanoscale.
o. Photoluminescence spectroscopy (PL): a technique that measures the light emitted by nanomaterials after they are

excited with light, providing information on the optical properties and electronic structure of materials.

p. Circular dichroism (CD) spectroscopy: a technique that measures the differential absorption of left- and right-circularly polarized light by chiral nanomaterials
q. Laser-induced breakdown spectroscopy (LIBS): Measures the emission of light from a plasma generated by a high-power laser pulse to determine the elemental composition of a sample.
r. Terahertz spectroscopy: Measures the absorption and transmission of terahertz radiation by a sample to determine its electronic and structural properties.
s. Near-infrared spectroscopy (NIRS): Measures the absorption and transmission of near-infrared radiation by a sample to determine its molecular structure and identify its chemical composition.
t. Cathodoluminescence spectroscopy: Measures the emission of light by a sample after being excited by a beam of electrons to determine its electronic and structural properties.
u. Electron energy loss spectroscopy (EELS): EELS is used to study the electronic structure and bonding of nanoparticles, providing high-resolution information on their chemical composition and bonding.
v. Time-resolved spectroscopy: Time-resolved spectroscopy is used to study the dynamics and kinetics of nanoparticle reactions, providing information on the behavior of nanoparticles over time.
w. Fluorescence spectroscopy: Fluorescence spectroscopy is used to study the emission of light from nanoparticles, and is commonly used to study the optical properties and behavior of fluorescent nanoparticles.
x. Nuclear magnetic resonance spectroscopy (NMR): NMR is a powerful technique used to study the electronic environment of atoms in nanomaterials by measuring their magnetic properties under an external magnetic field. It can provide information on the chemical structure,

composition, and dynamics of nanoparticles, as well as their interactions with other molecules or surfaces. NMR is particularly useful for studying the surface chemistry of nanomaterials and can help in the design and optimization of functionalized nanoparticles for various applications.

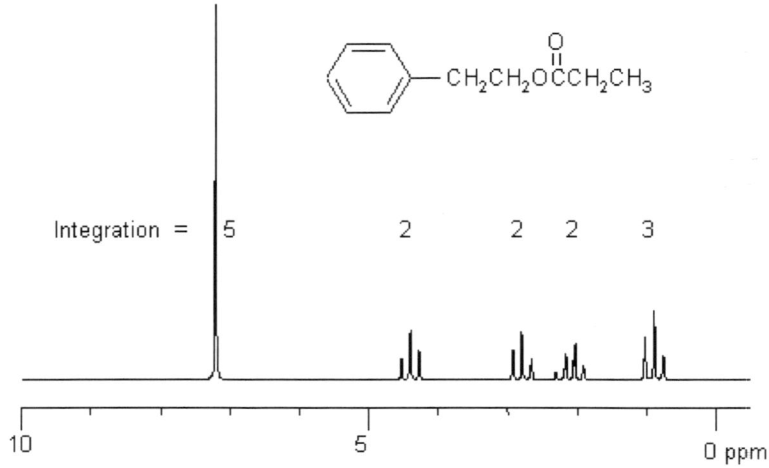

Fig 16. NMR spectrum

CHALLENGES

- What are some common techniques used for the characterization of nanomaterials, and how do they work?
- How do the properties of nanomaterials vary depending on their size, shape, composition, and structure?
- What are some challenges associated with the characterization of nanomaterials, and how can they be addressed?
- What is the role of electron microscopy in the analysis of nanomaterials, and how is it used to visualize and characterize particles?
- What are some advantages and limitations of transmission electron microscopy (TEM) and scanning electron microscopy (SEM) for the analysis of nanomaterials?
- How does atomic force microscopy (AFM) work, and what are some applications of this technique in the analysis of nanomaterials?
- What is the role of X-ray diffraction (XRD) in the analysis of nanomaterials, and how is it used to determine crystal structure and composition?
- How does X-ray photoelectron spectroscopy (XPS) work, and what information can be obtained from this technique in the analysis of nanomaterials?
- What is the role of Fourier transform infrared spectroscopy (FTIR) in the analysis of nanomaterials, and how is it used to identify chemical functional groups?

- How does Raman spectroscopy work, and what are some applications of this technique in the analysis of nanomaterials?
- What is the role of dynamic light scattering (DLS) in the analysis of nanomaterials, and how is it used to measure particle size and size distribution?
- What are some other techniques used for the characterization of nanomaterials, such as differential scanning calorimetry (DSC), thermogravimetric analysis (TGA), and zeta potential analysis?
- How do the properties of nanomaterials change in different environments, such as in biological systems or under varying temperature and pH conditions?
- What is the role of in situ and operando analysis in the study of nanomaterials, and how are these techniques used to understand reaction mechanisms?
- How can statistical analysis and modeling be used to interpret and compare data from different analytical techniques in the analysis of nanomaterials?
- What are some challenges associated with the quantification of nanomaterials, such as particle concentration and mass, and how can they be addressed?
- What is the role of multi-scale analysis in the study of nanomaterials, and how can insights from different length and time scales be integrated?
- How can the analysis of nanomaterials be used to understand and mitigate potential environmental and health risks associated with their use?

CHAPTER FIVE

5.1 Surface Chemistry of Nanoparticles

Surface chemistry of nanomaterials is the study of the chemical and physical phenomena that occur at the surface of nanomaterials. It involves understanding the structure and properties of the surface of nanomaterials, as well as the interactions of the surface with its environment. The surface of nanomaterials is of particular importance because it plays a crucial role in many applications of nanomaterials, including catalysis, sensing, and drug delivery. Surface chemistry of nanomaterials is a multidisciplinary field that draws on concepts from chemistry, physics, materials science, and biology to understand and manipulate the surface properties of nanomaterials.

5.2 Adsorption on Nanoparticle Surfaces

Adsorption is the process of accumulation of molecules, atoms, or ions on the surface of a solid material. It is an important phenomenon in various fields of science, such as surface chemistry, materials science, environmental science, and nanotechnology. In nanotechnology, adsorption plays a crucial role in the synthesis, characterization, and applications of nanoparticles.

Nanoparticles have a high surface area-to-volume ratio, which means that a large proportion of their atoms or molecules are present on the surface. This high surface area-to-volume ratio makes the surface of nanoparticles highly reactive and provides a large number of active sites for adsorption. The surface of nanoparticles can be modified to tune their adsorption properties, which can be utilized for various applications.

Adsorption on nanoparticle surfaces can be classified into two types: physical adsorption and chemical adsorption. Physical adsorption, also known as physisorption, occurs due to weak Van der Waals forces between the adsorbate molecules and the surface of the nanoparticle. This type of adsorption is reversible and can be affected by factors such as temperature, pressure, and the size of the adsorbate molecule. Physical adsorption is typically observed at low temperatures and is often used for gas separation and storage applications.

Chemical adsorption, also known as chemisorption, occurs due to the formation of chemical bonds between the adsorbate molecules and the surface of the nanoparticle. This type of adsorption is irreversible and requires the activation of the surface of the nanoparticle. The chemical bonding between the adsorbate molecules and the nanoparticle surface can be used for various applications such as catalysis, sensing, and drug delivery.

The surface chemistry of nanoparticles plays a crucial role in determining the adsorption properties of nanoparticles. The surface of nanoparticles can be modified to control the adsorption properties by introducing functional groups or coatings. For example, the surface of gold nanoparticles can be modified by attaching thiol groups, which provide a strong binding site for various biomolecules.

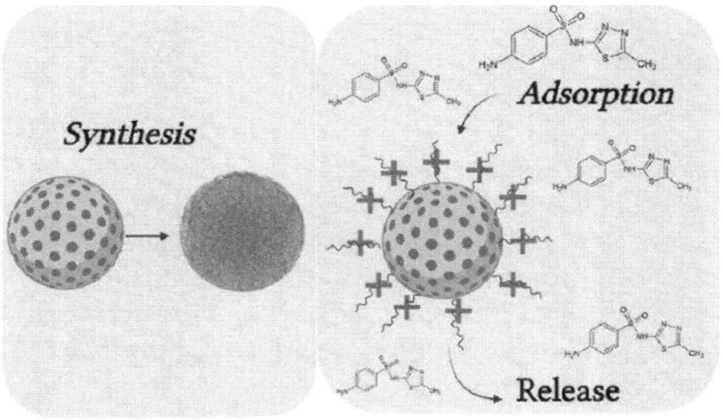

Fig 17. Adsorption of nanoparticles

In addition to the surface chemistry, the size and shape of nanoparticles can also affect the adsorption properties. As the size of nanoparticles decreases, the surface area-to-volume ratio increases, which results in a higher number of active sites for adsorption. Similarly, the shape of nanoparticles can affect the distribution of surface atoms and hence the adsorption properties. For example, cubic nanoparticles have more surface atoms than spherical nanoparticles, which results in a higher number of active sites for adsorption.

Various techniques can be used to study the adsorption properties of nanoparticles. Surface plasmon resonance spectroscopy is a commonly used technique for the real-time monitoring of adsorption on nanoparticle surfaces. This technique is based on the change in the refractive index of the surrounding medium due to the adsorption of molecules on the nanoparticle surface. X-ray photoelectron spectroscopy is another commonly used technique for the characterization of the chemical composition of nanoparticle surfaces. This technique involves the excitation of electrons from the surface of the nanoparticle using X-rays, which can provide information about the chemical bonds present on the nanoparticle surface.

In summary, adsorption on nanoparticle surfaces is an important phenomenon that plays a crucial role in the synthesis, characterization, and applications of nanoparticles. The surface chemistry, size, and shape of nanoparticles can affect the adsorption properties and can be used to tune the properties for various applications. Various techniques are available to study the adsorption properties of nanoparticles, which can provide valuable information for the design and optimization of nanoparticle-based materials and devices.

5.3 Surface Functionalization of Nanoparticles

Surface functionalization of nanoparticles refers to the process of modifying the surface properties of nanoparticles through the attachment of functional groups, such as organic molecules, polymers, or biomolecules. This process is crucial for controlling the surface chemistry, stability, and reactivity of nanoparticles, and plays a significant role in various applications, including drug delivery, imaging, and sensing.

The surface of nanoparticles is highly reactive and prone to agglomeration and degradation, which limits their use in practical applications. Surface functionalization provides a means to overcome these limitations by creating a stable and functional surface layer that can enhance the dispersibility and biocompatibility of nanoparticles.

Surface functionalization of nanoparticles can also be used to tailor the surface properties, such as hydrophobicity, hydrophilicity, or charge, to control the interaction of nanoparticles with biological systems. For example, hydrophobic functional groups, such as alkyl chains, can improve the uptake and penetration of nanoparticles into cells, while hydrophilic functional groups, such as polyethylene glycol (PEG), can enhance the biocompatibility and circulation time of nanoparticles in vivo.

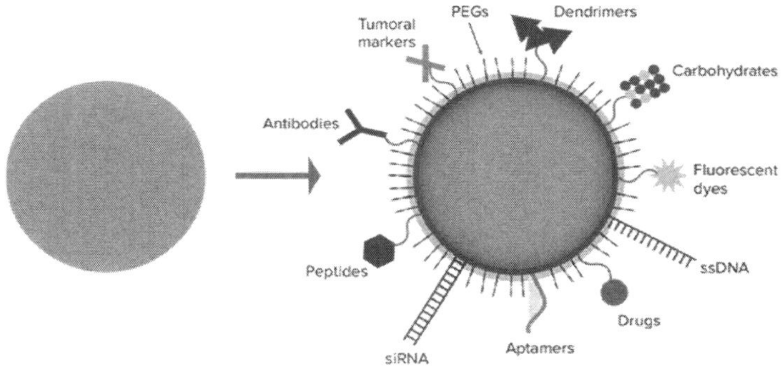

Fig 18. Surface functionalization of nanoparticles

There are several methods for surface functionalization of nanoparticles, some of which are:

Covalent functionalization: In this method, chemical bonds are formed between the surface of nanoparticles and functional groups through chemical reactions. This method provides strong bonding and stability to the surface of nanoparticles.

Non-covalent functionalization: In this method, functional groups are attached to the surface of nanoparticles through non-covalent interactions such as electrostatic interactions, hydrogen bonding, van der Waals forces, and hydrophobic interactions.

Polymer functionalization: In this method, a layer of polymer is coated on the surface of nanoparticles. The functional groups are then attached to the polymer layer. Polymer functionalization provides better stability and solubility to the nanoparticles.

Ligand exchange: In this method, the original ligands on the surface of nanoparticles are replaced with new functional ligands by a ligand exchange reaction. This method provides a controlled surface functionalization process with high reproducibility.

Physical adsorption: In this method, the functional groups are adsorbed on the surface of nanoparticles through physical adsorption. This method is simple, easy, and cost-effective. However, the adsorbed functional groups may not be strongly attached to the surface of nanoparticles.

Click chemistry: In this method, a specific chemical reaction is used to attach functional groups to the surface of nanoparticles. This method provides a high degree of control over the surface functionalization process and can be used for a wide range of functional groups.

Biofunctionalization: In this method, biomolecules such as proteins, enzymes, or DNA are attached to the surface of nanoparticles. This method is used in various biomedical applications such as drug delivery, imaging, and biosensors.

Overall, surface functionalization plays a crucial role in the development of nanomaterials with tailored properties and specific applications. The choice of functionalization method depends on the properties of the nanoparticles and the desired application.

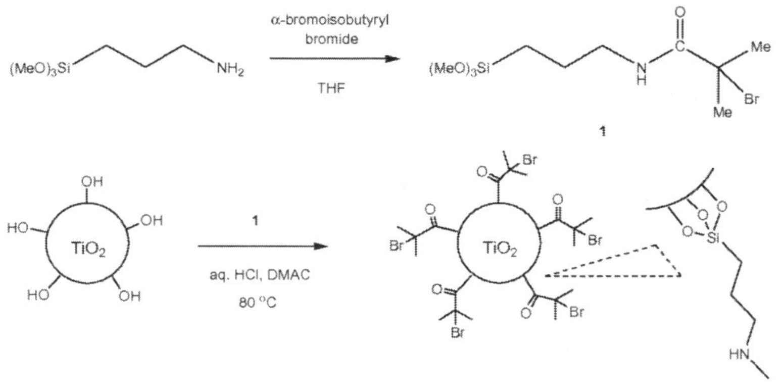

Surface modification of metal oxide nanoparticles

5.4 Self-Assembly of Nanoparticles

Self-assembly is a process by which individual particles spontaneously organize into well-defined patterns or structures without the intervention of external forces or templates. This process is driven by interparticle interactions such as van der Waals forces, hydrogen bonding, electrostatic forces, and hydrophobic interactions. The self-assembly of nanoparticles has received considerable attention due to the potential to create novel materials with unique properties and functions.

The self-assembly of nanoparticles has numerous potential applications in areas such as electronics, optics, and medicine. In electronics, self-assembled nanoparticle monolayers can be used to create nanoscale electronic devices such as transistors and sensors. In optics, the self-assembly of nanoparticles can be used to create photonic crystals with unique optical properties. In medicine, self-assembled nanoparticle structures can be used as drug delivery vehicles or for targeted imaging.

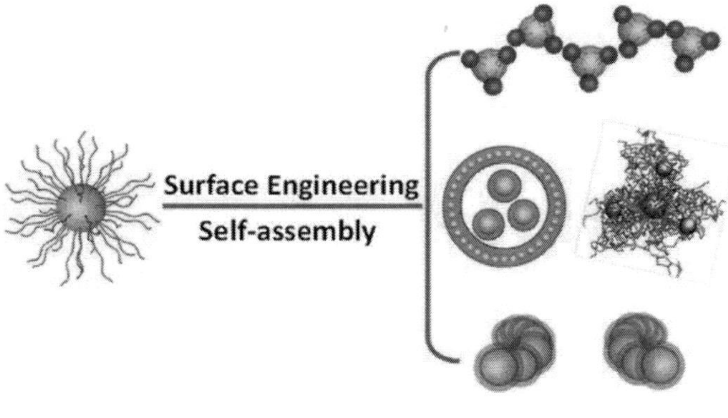

Fig 19. Self assembly of nanoparticles

There are several methods for self-assembly of nanoparticles, including:

a. Chemical methods: In this method, the surface of nanoparticles is modified by functional groups or surfactants, which help to direct the assembly process.
b. Physical methods: This method involves the use of external forces such as magnetic fields, electric fields, and shear forces to control the assembly process.
c. Templated methods: In this method, a template is used to guide the assembly process. The template can be a molecule, a substrate, or a patterned surface.
d. Biological methods: In this method, biomolecules such as proteins, DNA, and viruses are used to direct the assembly process.
e. Electrostatic methods: In this method, the self-assembly process is controlled by the interaction between charged nanoparticles.
f. Solvent evaporation: In this method, nanoparticles are dispersed in a solvent and the solvent is allowed to evaporate, leading to the self-assembly of the nanoparticles.

g. Layer-by-layer assembly: In this method, nanoparticles are alternately deposited onto a substrate, leading to the formation of a multilayer structure.
h. Colloidal crystal templating: In this method, a colloidal crystal is used as a template for the assembly of nanoparticles.
i. Langmuir-Blodgett technique: In this method, a monolayer of nanoparticles is deposited onto a substrate using the Langmuir-Blodgett technique.
j. Dip-pen nanolithography: In this method, a nanoscale tip is used to deposit nanoparticles onto a substrate, allowing for the precise control of the assembly process.

Overall, the self-assembly of nanoparticles is a promising approach for the creation of novel materials with unique properties and functions. Through the control of interparticle interactions and surface functionalization, well-defined structures can be created with potential applications in a wide range of fields.

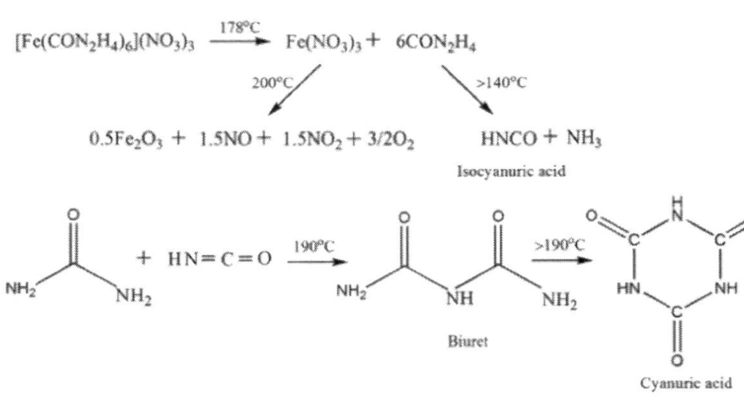

Synthesis of self-assembled prismatic iron oxide nanoparticles by a novel thermal decomposition route

CHALLENGES

- What is surface chemistry and why is it important in the field of nanoparticles?
- What is adsorption and how does it relate to surface chemistry?
- How can nanoparticles be functionalized to improve their properties?
- What are the different types of adsorption isotherms and how are they used in surface chemistry?
- How do surface properties such as surface charge and surface energy affect adsorption?
- What is Langmuir adsorption theory and how is it used to describe adsorption?
- How does the BET method work to measure surface area of nanoparticles?
- What is the difference between physical and chemical adsorption?
- What are the applications of surface functionalized nanoparticles in biomedical and environmental fields?
- How can surface functionalization of nanoparticles be achieved using various techniques?
- What is self-assembly of nanoparticles and how does it occur?
- What is the role of surfactants in the self-assembly of nanoparticles?

- How can self-assembly be controlled to create specific nanostructures?
- What are the advantages and disadvantages of self-assembled nanostructures?
- What is the future of surface chemistry in the field of nanotechnology?
- What are some common methods for characterizing the surface properties of nanoparticles?
- How do temperature, pressure, and solution pH affect adsorption behavior?
- What is the role of ligands in surface functionalization of nanoparticles?
- How can the size and shape of nanoparticles influence their adsorption properties?
- What are some challenges in the development of surface-functionalized nanoparticles for real-world applications?

CHAPTER SIX

6.1 Quantum Dots

Quantum dots (QDs) are nanoparticles with unique optical and electronic properties that arise due to quantum confinement effects. They are typically in the size range of 1-10 nanometers and are composed of elements such as cadmium, selenium, and zinc. QDs have found numerous applications in the fields of electronics, optoelectronics, and biomedicine due to their tunable bandgap, high quantum yield, and narrow emission spectra.

The optical and electronic properties of QDs are determined by their size and shape. As the size of the QD decreases, the bandgap energy increases, leading to a blue shift in their absorption and emission spectra. This size-dependent property is a result of the confinement of the electrons and holes within the nanocrystal, which causes quantization of their energy levels.

The ability to tune the emission spectra of QDs by controlling their size has led to their widespread use in biological imaging and labeling applications. QDs are often coated with biocompatible materials such as polymers or peptides to improve their stability and minimize their toxicity. These functionalized QDs have been used as fluorescent probes to visualize cellular processes and in vivo imaging of tissues and organs.

In addition to their optical properties, QDs also have unique electronic properties that make them promising candidates for applications in electronics and optoelectronics. The narrow emission spectra of QDs make them suitable for use in color displays, and their high quantum yield makes them useful for efficient light-emitting diodes (LEDs). QDs can also be used as sensitizers in solar cells, as they can absorb light over a broad range of wavelengths and convert it into electrical energy.

Overall, the unique optical and electronic properties of QDs make them a versatile and promising class of nanomaterials with potential applications in a wide range of fields. However, their potential toxicity and environmental impact must also be carefully considered as they become more widely used.

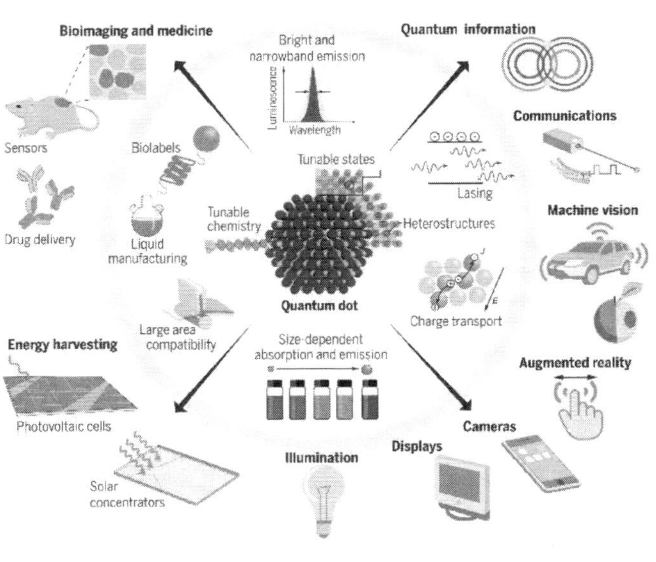

Fig 20. Quantum dots

6.2 Types of Quantum Dots

Quantum dots (QDs) can be categorized based on their composition, structure, and size. The following are some common types of quantum dots:

II-VI quantum dots: These QDs are made of a combination of group II and group VI elements in the periodic table, such as cadmium selenide (CdSe), cadmium telluride (CdTe), and zinc sulfide (ZnS).

III-V quantum dots: These QDs are made of a combination of group III and group V elements, such as indium arsenide (InAs), indium phosphide (InP), and gallium arsenide (GaAs).

Core-shell quantum dots: These QDs have a core of one material surrounded by a shell of another material. The shell is usually made of a material with a wider bandgap than the core, which provides better stability and enhanced optical properties. Examples include CdSe/ZnS and InP/ZnS QDs.

Alloyed quantum dots: These QDs are made of a combination of two or more different elements, such as cadmium zinc sulfide (CdZnS) or indium gallium arsenide (InGaAs).

Doped quantum dots: These QDs have impurities or dopants added to their structure to modify their electronic and optical properties. Examples include Mn-doped CdSe QDs, which exhibit magnetic properties, and Cu-doped InP QDs, which emit red light.

Perovskite quantum dots: These QDs are a relatively new type of QD made of a perovskite crystal structure, such as CsPbX3 (X = Cl, Br, I). They have shown great promise for use in optoelectronic devices due to their high quantum yield and tunable optical properties.

Organic-inorganic hybrid quantum dots: These QDs have an organic molecule attached to their surface, which can enhance their stability and biocompatibility. Examples include CdSe QDs functionalized with polyethylene glycol (PEG) and gold-coated QDs functionalized with a thiolated organic molecule.

Carbon quantum dots (CQDs) are a type of quantum dot made from carbon-based materials such as 73rapheme and carbon nanotubes. They possess unique properties such as high photoluminescence, biocompatibility, and low toxicity, making them suitable for various applications in optoelectronics, bioimaging, and sensors. CQDs can be synthesized using various methods such as chemical synthesis, microwave-assisted synthesis, and laser ablation

Overall, the type of quantum dot used depends on the specific application and desired properties, such as emission wavelength, stability, and biocompatibility.

6.3 Synthesis and Characterization of Quantum Dots

Quantum dots are nanoscale materials that have unique electronic and optical properties due to their small size and the quantum confinement effect. These materials have gained significant attention in recent years for their potential use in a wide range of applications, including biological imaging, photovoltaics, and optoelectronics. The synthesis and characterization of quantum dots are essential for understanding their properties and developing practical applications.

Synthesis of Quantum Dots:

There are several methods for synthesizing quantum dots, including colloidal synthesis, sol-gel synthesis, and vapor phase synthesis. Colloidal synthesis is the most widely used method for synthesizing quantum dots. In this method, metal precursors are dissolved in a solution of surfactants and heated under controlled conditions. As the solution cools, the metal ions nucleate and form small clusters that eventually grow into nanoscale particles. The size and shape of the resulting quantum dots can be controlled by adjusting the reaction conditions, such as the concentration of the precursor solution and the reaction temperature.

a. Chemical Precipitation Method: This is one of the most common methods for synthesizing QDs. It involves the precipitation of QDs from a solution containing metal salts and a reducing agent. The size and shape of the QDs can be controlled by varying the concentration of the reactants, the reaction temperature, and the reaction time.
b. Hydrothermal Synthesis: This method involves the synthesis of QDs using high-temperature and high-pressure conditions. The reactants are typically placed in an autoclave and heated to a high temperature for a period of time, allowing the QDs to form. This method is useful for synthesizing QDs with narrow size distributions and high crystallinity.
c. Sol-Gel Method: This method involves the formation of QDs from a sol-gel precursor solution. The precursor solution is typically a mixture of metal alkoxides, which are hydrolyzed and condensed to form a gel. The gel is then heated to a high temperature to form the QDs. This method allows for the formation of QDs with a high degree of control over their size and shape.
d. Microwave Synthesis: This method involves the use of microwave radiation to synthesize QDs. The reactants are

typically placed in a microwave reactor, which heats the reaction mixture rapidly and uniformly. This method allows for the rapid synthesis of QDs with high crystallinity and narrow size distributions.

e. Electrochemical Synthesis: This method involves the synthesis of QDs using an electrochemical cell. The QDs are formed at the electrode surface by the reduction of metal ions in the solution. This method allows for the synthesis of QDs with a high degree of control over their size and shape.

Characterization of Quantum Dots:

a. Optical Characterization: Optical characterization techniques are used to study the optical properties of QDs, including their absorption, emission, and fluorescence spectra. UV-Vis absorption spectroscopy, fluorescence spectroscopy, and time-resolved fluorescence spectroscopy are commonly used techniques for optical characterization of QDs.
b. Structural Characterization: Structural characterization techniques are used to study the structural properties of QDs, including their size, shape, and crystal structure. Transmission electron microscopy (TEM), scanning electron microscopy (SEM), X-ray diffraction (XRD), and atomic force microscopy (AFM) are commonly used techniques for structural characterization of QDs.
c. Surface Characterization: Surface characterization techniques are used to study the surface properties of QDs, including their surface charge, surface chemistry, and surface morphology. Fourier-transform infrared spectroscopy (FTIR), X-ray photoelectron spectroscopy (XPS), and dynamic light scattering (DLS) are commonly used techniques for surface characterization of QDs.

In summary, the synthesis and characterization of QDs are important for understanding their unique properties and for developing new applications. There are several methods for synthesizing QDs, each with its own advantages and limitations. Optical, structural, and surface characterization techniques are commonly used for the characterization of QDs.

6.4 Applications of Quantum Dots

Quantum dots have a wide range of applications due to their unique properties, such as size-tunable optical and electronic properties, high fluorescence quantum yield, and photochemical stability. Here are some of the applications of quantum dots:

1. Biological imaging: Quantum dots have been widely used as probes for in vivo and in vitro imaging in biological research due to their superior brightness and photostability. They can be used for imaging cells, tissues, and even whole organisms.
2. Drug delivery: Quantum dots have been explored as potential carriers for drug delivery due to their small size and high surface area. They can be functionalized with specific targeting ligands and loaded with drugs for targeted delivery to diseased cells.
3. Solar cells: Quantum dots can be used as light absorbers in solar cells due to their ability to efficiently convert light into electrical energy. They can be incorporated into the active layer of solar cells to improve their efficiency.
4. Lighting: Quantum dots have been used to produce high-quality lighting sources, such as light-emitting diodes (LEDs) and displays. They offer a wider range of colors and a higher color purity compared to traditional lighting sources.

5. Sensing: Quantum dots can be used as sensors due to their high sensitivity to changes in their environment. They can be functionalized with specific receptors to detect various analytes, such as heavy metals, gases, and biomolecules.
6. Data storage: Quantum dots have been explored as potential candidates for data storage due to their ability to store information in a quantum state. They can be used to develop high-density storage devices with high-speed read and write capabilities.
7. Biomedical applications: Quantum dots have been used for a variety of biomedical applications, such as tumor targeting, gene delivery, and photodynamic therapy. They offer unique advantages, such as high selectivity, sensitivity, and biocompatibility.
8. Security applications: Quantum dots have been explored as potential candidates for security applications, such as authentication and anti-counterfeiting. They can be used as unique identifiers due to their size-tunable optical properties and high photostability.

In summary, quantum dots have a wide range of applications in various fields, including biological imaging, drug delivery, solar cells, lighting, sensing, data storage, biomedical applications, and security applications. With ongoing research and development, more applications of quantum dots are likely to emerge in the future.

CHALLENGES

- What are quantum dots?
- How are quantum dots different from regular materials?
- What are the properties of quantum dots?
- What are the applications of quantum dots?
- How are quantum dots synthesized?
- What is the role of size in quantum dots?
- What are the types of quantum dots?
- How do quantum dots emit light?
- What is the quantum confinement effect?
- What is the photoluminescence of quantum dots?
- How do quantum dots enable high-resolution imaging?
- How do quantum dots enhance solar cell efficiency?
- How are quantum dots used in LED displays?
- What are the challenges of using quantum dots?
- What is the toxicity of quantum dots?
- How can the toxicity of quantum dots be minimized?
- What are the most commonly used characterization techniques for quantum dots?
- How does the size of quantum dots affect their optical and electronic properties?
- What are the potential applications of quantum dots in biomedicine?
- How can quantum dots be used in biosensors?
- What are the potential applications of quantum dots in energy production?
- How can quantum dots be used in photovoltaic devices?

- What are the potential applications of quantum dots in optoelectronic devices?
- How can quantum dots be used in light-emitting diodes (LEDs)?
- What is the role of quantum dots in quantum computing?
- How can quantum dots be used in data storage?
- How can quantum dots be used in chemical sensing?
- What is the role of quantum dots in catalysis?
- How can quantum dots be used in environmental monitoring?
- What are the potential challenges of using quantum dots in real-world applications?
- What are the potential ethical and safety concerns of using quantum dots?
- What is the future of quantum dots in the field of nanotechnology and beyond?

CHAPTER SEVEN

7.1 Nanorods and Nanowires

Nanorods and nanowires are one-dimensional (1D) nanostructures that possess length dimensions ranging from a few nanometers to several micrometers, and diameters typically ranging from 1 to 100 nanometers. These nanostructures are unique because their aspect ratio (length-to-diameter ratio) is much larger than that of conventional bulk materials, which imparts novel optical, electronic, and mechanical properties. Nanorods and nanowires have emerged as promising building blocks for the development of nanoelectronics, nanophotonics, and nanomechanical devices. They can be synthesized using a variety of methods, such as template-based synthesis, solution-phase synthesis, and vapor-phase synthesis, and can be made from various materials including metals, semiconductors, and oxides.

7.2 Synthesis and Characterization of Nanorods and Nanowires

Synthesis of Nanorods and Nanowires

There are several methods to synthesize nanorods and nanowires, including template-assisted synthesis, vapor-liquid-solid (VLS) method, solution-phase synthesis, and hydrothermal synthesis.

a. Template-Assisted Synthesis: Template-assisted synthesis is a method where the nanorods or nanowires are grown within a pre-formed template. The template can be a nanoporous membrane or anodic aluminum oxide (AAO). The pores of the template determine the dimensions of the nanorods or nanowires. The template is first filled with the precursor solution, and then the precursor is reduced or oxidized to form the nanorods or nanowires. The template is then removed by chemical or physical methods. This method can be used to synthesize a wide variety of materials.
b. Vapor-Liquid-Solid (VLS) Method: The VLS method involves a catalyst that is placed on a substrate, and the substrate is then heated in the presence of a gas-phase precursor. The catalyst melts and forms droplets, which catalyze the growth of the nanorods or nanowires from the gas-phase precursor. The size and shape of the nanorods or nanowires are controlled by the size and shape of the catalyst droplets.
c. Solution-Phase Synthesis: Solution-phase synthesis is a simple and effective method to synthesize nanorods and nanowires. The method involves the reduction of metal ions in a solution in the presence of a surfactant or capping agent. The surfactant or capping agent controls the size and shape of the nanorods or nanowires.
d. Hydrothermal Synthesis: Hydrothermal synthesis involves the growth of nanorods or nanowires in an aqueous solution at high temperature and pressure. The growth of

the nanorods or nanowires is controlled by the temperature, pressure, and pH of the solution.

Characterization of Nanorods and Nanowires:

There are several techniques to characterize the nanorods and nanowires, including transmission electron microscopy (TEM), scanning electron microscopy (SEM), X-ray diffraction (XRD), and Raman spectroscopy.

a. Transmission Electron Microscopy (TEM): TEM is a high-resolution imaging technique that can be used to determine the size, shape, and crystal structure of the nanorods or nanowires. TEM provides a direct measurement of the length and diameter of the nanorods or nanowires.
b. Scanning Electron Microscopy (SEM): SEM is a high-resolution imaging technique that can be used to determine the size, shape, and morphology of the nanorods or nanowires. SEM provides a top-down view of the nanorods or nanowires.
c. X-ray Diffraction (XRD): XRD is a technique that can be used to determine the crystal structure and orientation of the nanorods or nanowires. XRD provides information about the crystal structure and size of the nanorods or nanowires.
d. Raman Spectroscopy: Raman spectroscopy is a technique that can be used to determine the chemical composition and structure of the nanorods or nanowires. Raman spectroscopy provides information about the vibrational modes of the nanorods or nanowires.

Fig 21. SEM image of Nanowires

7.3 Applications of Nanorods and Nanowires

Nanorods and nanowires are promising nanomaterials that find applications in a wide range of fields, including electronics, photonics, catalysis, energy, and biomedical engineering. Some of the most significant applications of nanorods and nanowires are:

a. Electronics and Photonics: Nanorods and nanowires can be used as building blocks for electronic and photonic devices. They can be integrated into circuits to form transistors, diodes, and other electronic components. Additionally, their unique optical properties make them suitable for applications in lasers, photodetectors, and sensors.
b. Catalysis: Nanorods and nanowires are excellent catalysts due to their high surface area-to-volume ratio and unique surface properties. They are used in a wide range of catalytic reactions, including hydrogenation, oxidation, and carbon dioxide reduction.
c. Energy: Nanorods and nanowires have shown potential in energy applications, such as solar cells and batteries. For example, nanowires can be used as the active layer in solar cells to increase their efficiency, while nanorods can be

used as electrodes in lithium-ion batteries to improve their performance.

d. Biomedical Engineering: Nanorods and nanowires are being explored for use in biomedical engineering applications, such as drug delivery, imaging, and biosensors. They can be functionalized with various biomolecules to target specific cells or tissues in the body.
e. Environmental Remediation: Nanorods and nanowires can be used for environmental remediation, such as water purification and air filtration. For example, nanowires can be used to remove heavy metals and other pollutants from water, while nanorods can be used as photocatalysts to degrade organic pollutants.
f. Flexible Electronics: Nanorods and nanowires can be integrated into flexible electronic devices due to their high flexibility and mechanical strength. They are being explored for use in applications such as flexible displays, wearable electronics, and electronic textiles.
g. Thermoelectric Devices: Nanorods and nanowires have shown promise in thermoelectric devices, which convert heat into electrical energy. They have high thermoelectric efficiency due to their unique electronic properties and small size.

Overall, the unique properties of nanorods and nanowires make them attractive materials for a wide range of applications in various fields. Ongoing research is expected to uncover new and innovative applications for these promising nanomaterials.

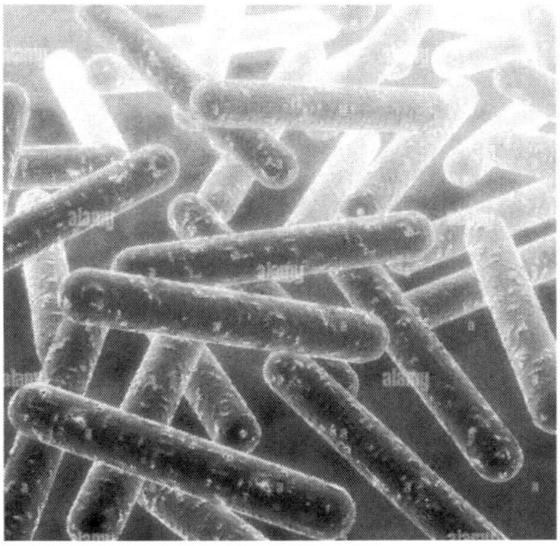

Fig 22. Nanorods

CHALLENGES

- What are nanorods and nanowires?
- How are nanorods and nanowires synthesized?
- What are the different synthesis methods for nanorods and nanowires?
- How is the size and aspect ratio of nanorods and nanowires controlled during synthesis?
- What are the commonly used materials for nanorod and nanowire synthesis?
- What are the most commonly used characterization techniques for nanorods and nanowires?
- How does the size and aspect ratio of nanorods and nanowires affect their properties?
- What are the potential applications of nanorods and nanowires in electronics?
- How can nanorods and nanowires be used in sensors?
- How can nanorods and nanowires be used in photovoltaic devices?
- What are the potential applications of nanorods and nanowires in energy storage?
- How can nanorods and nanowires be used in catalysis?
- What is the role of nanorods and nanowires in biomedicine?
- How can nanorods and nanowires be used in drug delivery?
- What are the potential challenges of using nanorods and nanowires in real-world applications?
- What are the potential ethical and safety concerns of using nanorods and nanowires?
- What is the future of nanorods and nanowires in the field of nanotechnology and beyond?

- How do nanorods and nanowires compare to other nanomaterials in terms of their properties and applications?
- How do environmental factors affect the properties and behavior of nanorods and nanowires?
- What are the current trends in research and development of nanorods and nanowires?

CHAPTER EIGHT

8.1 Nanofilms and nanoplates

Nanofilms and nanoplates are two-dimensional nanostructures that have attracted significant attention due to their unique electronic, optical, and mechanical properties. Nanofilms are thin films with a thickness of a few nanometers to several micrometers, whereas nanoplates are two-dimensional sheets with a thickness of a few nanometers to tens of nanometers and lateral dimensions ranging from a few nanometers to several micrometers

8.2 Synthesis and characterization of nanofilms and nanoplates

Synthesis of nanofilms and nanoplates

Nanofilms and nanoplates can be synthesized through various methods, some of which include:

a. Chemical Vapor Deposition (CVD): This method involves the reaction of gases on a heated substrate, resulting in the deposition of a thin film. The substrate is heated to a high temperature, and the gases are introduced in a controlled manner, leading to the formation of a thin film of the desired material.

b. Physical Vapor Deposition (PVD): This method involves the evaporation or sputtering of a material in a vacuum chamber. The vapors or particles from the material are deposited on a substrate, resulting in the formation of a thin film.
c. Sol-gel Method: This method involves the hydrolysis and condensation of metal alkoxides in a solution to form a gel. The gel is then heated to remove the solvent, resulting in the formation of a solid film.
d. Electrodeposition: This method involves the use of an electric current to deposit material on a substrate. The substrate is placed in an electrolyte solution, and an electric current is passed through it, resulting in the deposition of material on the substrate.
e. Langmuir-Blodgett Technique: This method involves the transfer of a monolayer of material from a water-air interface to a solid substrate. The monolayer is compressed at the water-air interface and then transferred to the substrate in a controlled manner, resulting in the formation of a thin film.
f. Dip Coating: This method involves the dipping of a substrate into a solution containing the material to be deposited. The substrate is then slowly withdrawn from the solution, resulting in the formation of a thin film on the substrate.
g. Spin Coating: This method involves the deposition of a material on a substrate by spinning the substrate at a high speed while a solution containing the material is added dropwise to the center of the substrate. The centrifugal force causes the solution to spread over the substrate, resulting in the formation of a thin film.

Overall, the choice of synthesis method for nanofilms and nanoplates depends on factors such as the desired material, the required film thickness, and the substrate material. Characterization techniques such as SEM, TEM, and AFM can be used to analyze the morphology and structure of the synthesized films.

Characterization of Nanofilms and Nanoplates

Nanofilms and nanoplates are thin film-like nanostructures with thicknesses ranging from a few nanometers to a few micrometers. Their properties can differ significantly from bulk materials due to quantum confinement effects, surface effects, and increased surface-to-volume ratios. To fully understand and optimize their properties, it is crucial to characterize them using various analytical techniques.

Some common characterization techniques for nanofilms and nanoplates are:

a. X-ray diffraction (XRD): XRD can be used to determine the crystallinity, crystal structure, and lattice parameters of nanofilms and nanoplates. The diffraction patterns can provide information about the crystallographic orientation, grain size, and texture.
b. Scanning electron microscopy (SEM): SEM is used to obtain high-resolution images of the surface morphology, size, and shape of nanofilms and nanoplates. SEM can also be combined with energy dispersive X-ray spectroscopy (EDS) to identify the elemental composition and distribution of the films.
c. Atomic force microscopy (AFM): AFM is a powerful technique for characterizing the topography, roughness, and mechanical properties of nanofilms and nanoplates. It uses a sharp tip to scan the surface of the sample and measures the force between the tip and the surface.
d. Transmission electron microscopy (TEM): TEM is used to obtain high-resolution images of the internal structure of

nanofilms and nanoplates. It can provide information about the crystal structure, grain boundaries, defects, and interfaces.

e. Spectroscopy techniques: Spectroscopic techniques such as Raman spectroscopy, infrared spectroscopy (IR), and UV-Vis spectroscopy can be used to study the chemical composition, electronic structure, and optical properties of nanofilms and nanoplates.

f. Ellipsometry: Ellipsometry is a non-destructive technique used to measure the thickness and optical properties of thin films, including nanofilms and nanoplates. It can provide information about the refractive index, extinction coefficient, and surface roughness.

g. X-ray photoelectron spectroscopy (XPS): XPS is a surface-sensitive technique that can be used to analyze the chemical composition and electronic structure of nanofilms and nanoplates. It can provide information about the chemical bonding, oxidation state, and surface chemistry of the films.

Overall, these characterization techniques can provide valuable insights into the properties and behavior of nanofilms and nanoplates, allowing for the optimization of their properties and the development of new applications.

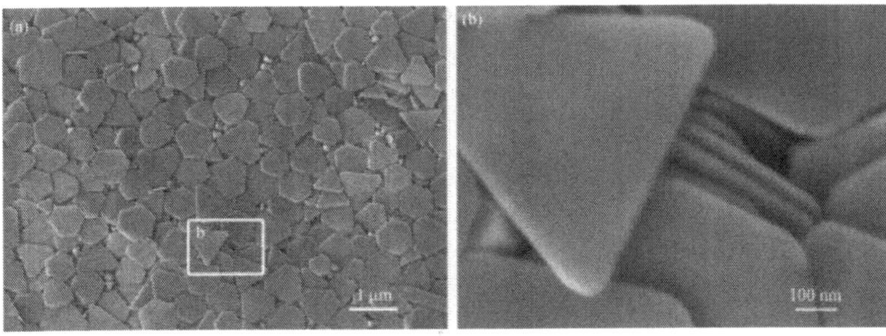

Fig 23. Nanoplates

8.3 Chemistry of Nanofilms and Nanoplates

Nanofilms and nanoplates are thin layers of material with thicknesses on the order of nanometers. These materials have unique properties that make them useful in a variety of applications, from electronics to energy storage. The chemistry of these materials is an important factor in determining their properties and applications.

Nanofilms are thin layers of material that can be deposited onto a surface using a variety of techniques, such as chemical vapor deposition (CVD), physical vapor deposition (PVD), and electrochemical deposition. Nanofilms can be made from a wide range of materials, including metals, semiconductors, and insulators.

The chemical composition of nanofilms is important in determining their properties. For example, the conductivity of a metal nanofilm is dependent on the metal used, while the bandgap of a semiconductor nanofilm is dependent on the semiconductor material used. In addition to the material composition, the structure and morphology of the nanofilm also play a role in determining its properties. For example, the thickness and grain size of a nanofilm can affect its mechanical properties, such as its strength and ductility.

Nanoplates are thin, two-dimensional materials with high aspect ratios. They can be synthesized using a variety of techniques, including solution-phase synthesis, hydrothermal synthesis, and chemical vapor deposition. Like nanofilms, the chemical composition of nanoplates is an important factor in determining their properties and applications.

One important class of nanoplates is graphene, a two-dimensional material composed of a single layer of carbon atoms arranged in a hexagonal lattice. Graphene has a number of unique properties, including high electrical conductivity, high mechanical strength, and high surface area. These properties make graphene useful in a variety of applications, including electronics, energy storage, and catalysis.

The chemistry of graphene is important in understanding its properties and applications. Graphene is a highly reactive material due to its large surface area, which makes it prone to chemical reactions with other materials. For example, graphene can react with oxygen to form graphene oxide, a material with a wide range of applications, including as a nanofiller in composite materials, as a catalyst, and as a biosensor.

Another important class of nanoplates is transition metal dichalcogenides (TMDs), which are composed of layers of transition metal atoms sandwiched between layers of chalcogen atoms. TMDs have a number of unique properties, including semiconducting behavior, high mechanical strength, and high surface area. These properties make TMDs useful in a variety of applications, including electronics and optoelectronics.

The chemistry of TMDs is important in determining their properties and applications. The electronic properties of TMDs are dependent on the number of layers and the type of transition metal and chalcogen used. For example, a single-layer TMD typically has a direct bandgap, making it useful in optoelectronic applications. In contrast, a multilayer TMD typically has an indirect bandgap, making it useful in electronic applications.

In addition to the material composition and structure, the surface chemistry of nanofilms and nanoplates is also important in determining their properties and applications.

The surface chemistry can affect the reactivity of the material, as well as its interactions with other materials. For example, functionalizing the surface of a nanofilm or nanoplate with specific chemical groups can improve its performance in a particular application.

In conclusion, the chemistry of nanofilms and nanoplates is an important factor in determining their properties and applications. The material composition, structure, and surface chemistry all play a role in determining the behavior of these materials, and understanding their chemistry is essential for developing new applications and improving existing ones.

8.4 Application of Nanofilms and Nanoplates

Nanofilms and nanoplates have a wide range of applications due to their unique properties, including high surface area, high aspect ratio, and tunable physical and chemical properties. In this article, we will discuss some of the applications of these materials in various fields.

a. Electronics: Nanofilms and nanoplates have revolutionized the field of electronics due to their high conductivity and tunable electronic properties. They are used in various electronic devices, such as transistors, sensors, and memory devices. For example, graphene, a two-dimensional material with high electrical conductivity, is used in the development of flexible electronic devices, such as touch screens and wearable sensors. Similarly, transition metal dichalcogenides (TMDs) are used in the development of high-performance field-effect transistors (FETs) due to their excellent electronic properties.

b. Energy Storage: Nanofilms and nanoplates are also used in the development of energy storage devices, such as batteries and supercapacitors. Due to their high surface area, these materials can store a large amount of charge in a small volume, resulting in high energy density. Graphene, for example, is used in the development of high-performance supercapacitors due to its high surface area and excellent electrical conductivity.
c. Catalysis: Nanofilms and nanoplates are also used as catalysts due to their high surface area and tunable chemical properties. They are used in various catalytic reactions, such as hydrogenation, oxidation, and reduction. For example, TMDs have been used as catalysts in the hydrogen evolution reaction (HER) due to their excellent catalytic activity.
d. Biomedical Applications: Nanofilms and nanoplates have also found applications in the field of medicine due to their biocompatibility and tunable surface properties. They are used in various biomedical applications, such as drug delivery, tissue engineering, and biosensors. For example, graphene oxide has been used as a drug delivery vehicle due to its ability to interact with various biomolecules.
e. Environmental Applications: Nanofilms and nanoplates are also used in environmental applications, such as water purification and air filtration. They are used as adsorbents due to their high surface area and tunable surface properties. For example, graphene oxide has been used as an adsorbent for heavy metal ions in water due to its high adsorption capacity.
f. Optoelectronics: Nanofilms and nanoplates are also used in the field of optoelectronics due to their excellent optical properties. They are used in various optoelectronic devices, such as solar cells, photodetectors, and light-emitting diodes (LEDs). For example, TMDs have been

used as active materials in photodetectors and LEDs due to their excellent optical properties.

g. Coatings: Nanofilms and nanoplates are also used as coatings due to their excellent mechanical properties and corrosion resistance. They are used to protect surfaces from wear and corrosion in various applications, such as automotive and aerospace industries. For example, graphene has been used as a coating material for various metallic surfaces due to its excellent corrosion resistance.

In conclusion, nanofilms and nanoplates have a wide range of applications in various fields, including electronics, energy storage, catalysis, biomedical applications, environmental applications, optoelectronics, and coatings. Their unique properties, including high surface area, tunable electronic and chemical properties, and excellent mechanical properties, make them promising candidates for various applications.

CHALLENGES

- What are nanoplates and nanofilms?
- How are nanoplates and nanofilms synthesized?
- What are the different synthesis methods for nanoplates and nanofilms?
- How is the thickness of nanoplates and nanofilms controlled during synthesis?
- What are the commonly used materials for nanoplate and nanofilm synthesis?
- What are the most commonly used characterization techniques for nanoplates and nanofilms?
- How does the thickness of nanoplates and nanofilms affect their properties?
- What are the potential applications of nanoplates and nanofilms in electronics?
- How can nanoplates and nanofilms be used in sensors?
- How can nanoplates and nanofilms be used in photovoltaic devices?
- What are the potential applications of nanoplates and nanofilms in energy storage?
- How can nanoplates and nanofilms be used in catalysis?
- What is the role of nanoplates and nanofilms in biomedicine?
- How can nanoplates and nanofilms be used in drug delivery?
- What are the potential challenges of using nanoplates and nanofilms in real-world applications?
- What are the potential ethical and safety concerns of using nanoplates and nanofilms?

- What is the future of nanoplates and nanofilms in the field of nanotechnology and beyond?
- How do nanoplates and nanofilms compare to other nanomaterials in terms of their properties and applications?
- How do environmental factors affect the properties and behavior of nanoplates and nanofilms?
- What are the current trends in research and development of nanoplates and nanofilms?

CHAPTER NINE

9.1 Nanoporous Materials

Nanoporous materials are materials that have a network of pores or channels that are typically less than 100 nanometers in size. These pores are usually empty spaces or filled with guest molecules. The high surface area and unique pore structure of nanoporous materials make them useful in various applications, such as catalysis, gas separation, and sensing. The synthesis of nanoporous materials involves various methods, including templating, self-assembly, and sol-gel methods. The properties of nanoporous materials can be tuned by controlling the size, shape, and connectivity of the pores.

9.2 Synthesis and Characterization of Nanoporous Materials

Synthesis of Nanoporous Materials

The synthesis of nanoporous materials can be achieved by several methods, including templating, sol-gel, and self-assembly. The choice of the synthesis method depends on the desired pore size, pore shape, and the chemical and physical properties of the material.

Templating method: This method involves using a template to create the desired pore size and shape. The template can be a hard template or a soft template. In the hard template method, the template is a solid material, and the nanoporous material is synthesized by depositing the desired material onto the template surface. The template is then removed to leave behind the nanoporous material. In the soft template method, the template is a surfactant or a polymer, and the nanoporous material is synthesized by assembling the material around the template. The template is then removed by either thermal treatment or chemical treatment.

Sol-gel method: The sol-gel method involves the synthesis of a gel-like material from a precursor solution. The precursor solution is typically a metal alkoxide, which undergoes hydrolysis and condensation reactions to form a gel. The gel is then subjected to heat treatment to remove the solvent and organic components, leaving behind a nanoporous material.

Self-assembly method: This method involves the spontaneous assembly of molecules into a nanoporous structure. The self-assembly can be driven by several mechanisms, including hydrogen bonding, van der Waals forces, and electrostatic interactions. The self-assembled structure can be stabilized by crosslinking or thermal treatment.

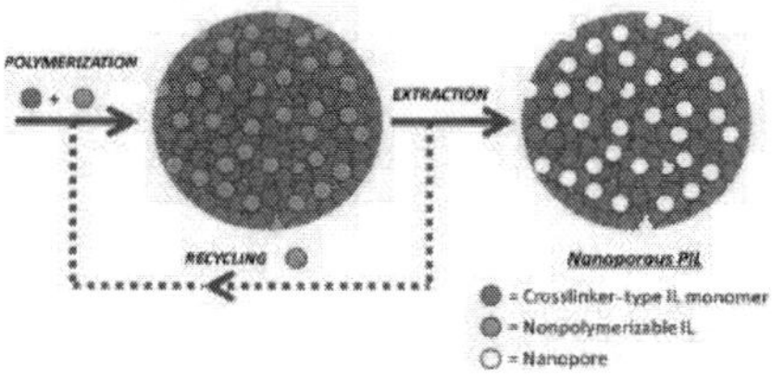

Fig 24. Synthesis of nanoporous materials

Characterization of Nanoporous Materials

The characterization of nanoporous materials involves the determination of pore size, pore shape, and pore volume. Several techniques can be used to characterize nanoporous materials, including gas adsorption, mercury intrusion porosimetry, and electron microscopy.

Gas adsorption: Gas adsorption involves exposing the nanoporous material to a gas and measuring the amount of gas adsorbed as a function of pressure. The pore size and pore volume can be determined from the adsorption isotherm.

Mercury intrusion porosimetry: Mercury intrusion porosimetry involves forcing mercury into the nanoporous material under high pressure and measuring the amount of mercury intruded as a function of pressure. The pore size and pore volume can be determined from the intrusion curve.

Electron microscopy: Electron microscopy can bused to visualize the nanoporous structure of the material. Transmission electron microscopy (TEM) can provide high-resolution images of the nanoporous structure, while scanning electron microscopy (SEM) can provide images of the surface morphology.

9.3 Chemistry of Nanoporous Materials

There are several types of nanoporous materials, including zeolites, mesoporous materials, metal-organic frameworks (MOFs), and covalent organic frameworks (COFs). Zeolites are crystalline materials with a three-dimensional network of nanoporous channels that are formed by the assembly of tetrahedral units. Mesoporous materials, on the other hand, have larger pores than zeolites and are typically synthesized using surfactant templates. MOFs and COFs are organic-inorganic hybrid materials with a crystalline structure and nanoporous channels.

The surface chemistry of nanoporous materials plays a crucial role in their properties and applications. The surface of nanoporous materials can be modified by functionalizing them with various organic and inorganic molecules. Functionalization can change the surface chemistry of the nanoporous material and can enhance its properties, such as its adsorption capacity, selectivity, and catalytic activity.

Nanoporous materials have a high surface area and contain pores with sizes in the range of a few nanometers to several micrometers. These materials have unique adsorption properties that make them useful in various applications such as gas storage, separation, catalysis, and sensors.

The adsorption properties of nanoporous materials are determined by several factors, including the pore size, pore shape, surface chemistry, and surface area. The pore size and shape affect the types of molecules that can be adsorbed, while the surface chemistry determines the strength of the interaction between the molecules and the surface.

The high surface area of nanoporous materials allows them to adsorb large amounts of gas or liquid molecules, making them suitable for applications such as gas storage and separation. For example, nanoporous materials such as metal-organic frameworks (MOFs) and zeolites are used for gas storage and separation because of their high surface area and selective adsorption properties.

9.4 Applications of Nanoporous Materials

Nanoporous materials find applications in various fields due to their unique properties. Some of the applications of nanoporous materials are:

a. Catalysis: Nanoporous materials have a high surface area and pore volume, making them an ideal catalyst support. The pore size and surface chemistry of the material can be tailored to fit the specific catalytic reaction.
b. Gas separation: The nanoporous structure of materials such as zeolites and metal-organic frameworks (MOFs) can be used for gas separation. The size and shape of the pores can be tuned to allow for the separation of different gases based on their molecular size and shape.
c. Energy storage: Nanoporous materials can be used for energy storage applications such as batteries and supercapacitors. The high surface area and pore volume allow for a large amount of energy to be stored.
d. Environmental applications: Nanoporous materials can be used for environmental applications such as water purification and air filtration. The nanoporous structure can be tailored to remove specific contaminants from water and air.
e. Drug delivery: The high surface area and pore volume of nanoporous materials make them ideal for drug delivery applications. The pore size can be tuned to allow for the controlled release of drugs over a specific time period.
f. Sensing: Nanoporous materials can be used for sensing applications such as gas sensing and biosensing. The high surface area and pore volume allow for the detection of low concentrations of analytes.
g. Optical applications: The nanoporous structure of materials such as photonic crystals can be used for optical applications such as sensing and imaging. The pore size can

be tuned to allow for the selective reflection and transmission of certain wavelengths of light.

h. Thermal insulation: Nanoporous materials can be used for thermal insulation applications due to their low thermal conductivity. The pore size and structure can be tailored to achieve the desired level of insulation.
i. Sound absorption: The nanoporous structure of materials such as aerogels can be used for sound absorption applications. The high surface area and pore volume allow for the absorption of sound waves.
j. Microelectronics: Nanoporous materials can be used for microelectronic applications such as sensors and electronic devices. The pore size and structure can be tailored to achieve the desired electronic properties.

Overall, the unique properties of nanoporous materials make them useful in a wide range of applications, from catalysis to microelectronics. As the field of nanotechnology continues to develop, new applications for nanoporous materials are likely to emerge.

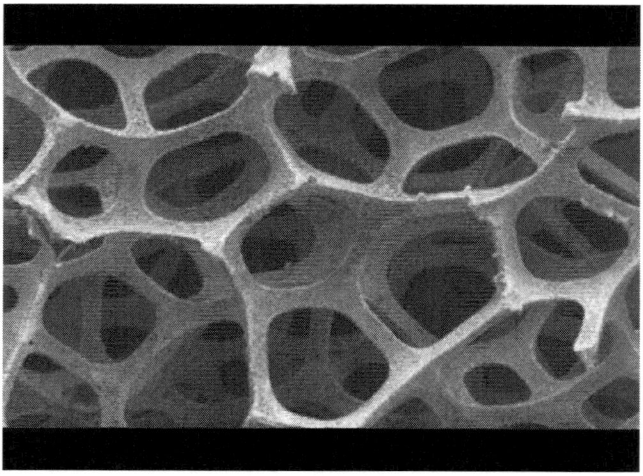

Fig 25. Nanoporous

CHALLENGES

- What are nanoporous materials?
- How are nanoporous materials synthesized?
- What are the different synthesis methods for nanoporous materials?
- How is the porosity of nanoporous materials controlled during synthesis?
- What are the commonly used materials for nanoporous material synthesis?
- What are the most commonly used characterization techniques for nanoporous materials?
- How does the porosity of nanoporous materials affect their properties?
- What are the potential applications of nanoporous materials in gas separation and purification?
- How can nanoporous materials be used in catalysis?
- How can nanoporous materials be used in sensors?
- What are the potential applications of nanoporous materials in drug delivery?
- How can nanoporous materials be used in energy storage?
- What is the role of nanoporous materials in biomedicine?
- What are the potential challenges of using nanoporous materials in real-world applications?

CHAPTER TEN

10.1 Nanocatalysis

Nanocatalysis is a rapidly growing field of research that involves the use of nanomaterials as catalysts for chemical reactions. A catalyst is a substance that facilitates a chemical reaction without being consumed in the reaction. Nanocatalysis takes advantage of the unique properties of nanoparticles, such as their high surface area, reactivity, and size-dependent electronic and optical properties, to enhance the efficiency and selectivity of chemical reactions.

One of the key advantages of nanocatalysis is that it enables chemical reactions to be carried out under milder conditions, such as lower temperatures and pressures, which can reduce energy consumption and minimize environmental impact. Nanocatalysts can also facilitate reactions that are difficult or impossible to achieve using conventional catalysts, leading to new opportunities for the synthesis of high-value chemicals and materials.

Nanocatalysts can be made from a variety of materials, including metals, metal oxides, and carbon-based materials such as graphene and carbon nanotubes. These materials can be synthesized using various methods such as chemical vapor deposition, sol-gel methods, and hydrothermal synthesis.

One of the most important factors in designing an effective nanocatalyst is to control the size, shape, and composition of the nanoparticles. For example, the size of the nanoparticles can affect their catalytic activity and selectivity, with smaller nanoparticles generally exhibiting higher activity due to their increased surface area. The shape of the nanoparticles can also influence their properties, with certain shapes such as cubes or octahedrons exhibiting enhanced activity due to their exposed facets.

Nanocatalysts can be used in a wide range of applications, including energy production, environmental remediation, and the synthesis of chemicals and materials. For example, nanocatalysts are used in fuel cells to facilitate the conversion of hydrogen and oxygen into water, producing electrical energy in the process. Nanocatalysts can also be used to convert carbon dioxide into useful chemicals, such as methanol, which can be used as a fuel or a feedstock for the production of other chemicals.

Another important application of nanocatalysis is in the synthesis of high-value chemicals and materials. Nanocatalysts can facilitate reactions that are difficult or impossible to achieve using conventional catalysts, leading to the production of unique materials with enhanced properties. For example, nanocatalysts can be used to synthesize metal-organic frameworks (MOFs), a class of materials that have high surface area, porosity, and tunable chemical properties, making them useful in gas storage, separation, and sensing applications.

Nanocatalysis is also being used to develop new methods for environmental remediation, such as the removal of pollutants from water and air. Nanocatalysts can facilitate the degradation of organic pollutants, such as pesticides and dyes, through processes such as photocatalysis and electrocatalysis.

However, the use of nanocatalysts also presents some challenges and potential risks. One of the main concerns is the potential toxicity of some nanomaterials, which can be harmful to human health and the environment. It is important to carefully evaluate the risks and benefits of using nanocatalysts and to develop appropriate safety protocols to minimize potential risks.

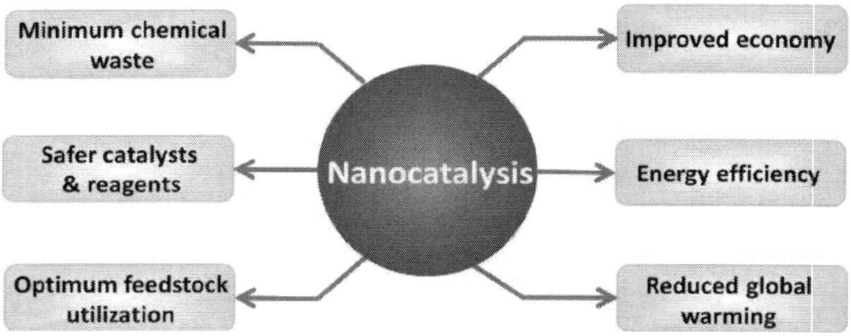

Fig 26. Overview of nanocatalysts

10.2 Catalytic Mechanisms of Nanoparticles

The catalytic mechanisms of nanoparticles depend on a variety of factors, including the size, shape, composition, and surface properties of the nanoparticles, as well as the nature of the chemical reaction being catalyzed. However, there are several general mechanisms that are commonly observed in nanocatalysis:

Adsorption and desorption: One of the key mechanisms by which nanoparticles catalyze chemical reactions is through the adsorption of reactant molecules onto their surfaces.

Nanoparticles have high surface areas, which provide a large number of sites for reactant molecules to adsorb. Once adsorbed, the reactant molecules can react with each other more easily, leading to the formation of product molecules. The product molecules can then desorb from the surface of the nanoparticle, allowing the catalyst to continue catalyzing the reaction.

Surface reactions: Nanoparticles can also catalyze reactions through surface reactions, which involve the reaction of molecules directly on the surface of the nanoparticle. In some cases, the nanoparticle itself can participate in the reaction, either by donating or accepting electrons or by serving as a template for the reaction.

Size and shape effects: The size and shape of nanoparticles can also have a significant impact on their catalytic activity. For example, smaller nanoparticles tend to have higher catalytic activity due to their larger surface area, while certain shapes, such as cubes or octahedrons, can exhibit enhanced activity due to their exposed facets.

Synergistic effects: In some cases, the catalytic activity of nanoparticles can be enhanced through synergistic effects between different components of the nanoparticle. For example, bimetallic nanoparticles, which consist of two different metals, can exhibit enhanced activity due to the interaction between the two metals.

Electron transfer: Nanoparticles can also facilitate chemical reactions through electron transfer processes, in which the nanoparticle donates or accepts electrons to or from reactant molecules. This can result in the formation of reactive intermediates or the activation of otherwise inert molecules.

Overall, the catalytic mechanisms of nanoparticles are complex and depend on a variety of factors. However, the unique properties of nanoparticles, such as their high surface area, size-dependent electronic and optical properties, and synergistic effects, make them highly effective catalysts for a wide range of chemical reactions.

10.3 Photocatalytic properties of Nanomaterials

Nanomaterials and photocatalysis are two areas of research that have gained significant attention in recent years due to their potential for addressing various environmental and energy challenges. Nanomaterials, which are materials with at least one dimension in the nanometer range (1-100 nm), exhibit unique properties that make them highly attractive for use in photocatalysis, a process that utilizes light to activate catalysts for promoting chemical reactions. In this article, we will delve into the exciting field of nanomaterials photocatalysis, covering the principles, applications, advantages, challenges, and future prospects of this cutting-edge research area.

Photocatalysis is a process that involves the use of a catalyst, which is a substance that facilitates a chemical reaction without being consumed, to promote a reaction upon exposure to light. When a photocatalyst is irradiated with light, it absorbs the energy from the photons, leading to the generation of electron-hole pairs. These electron-hole pairs can then participate in various chemical reactions, such as oxidation, reduction, and degradation of pollutants, on the surface of the photocatalyst. This unique property of photocatalysis makes it highly attractive for a wide range of applications, including environmental remediation, water purification, air purification, energy conversion, and even in biomedical applications.

Nanomaterials have shown tremendous potential as photocatalysts due to their unique properties arising from their nanoscale dimensions. One of the key advantages of using nanomaterials in photocatalysis is their high surface area-to-volume ratio, which provides more active sites for catalytic reactions compared to bulk materials. The increased surface area allows for enhanced absorption of light, leading to higher photocatalytic activity. Additionally, the small size of nanomaterials can also result in quantum confinement effects, where the electronic properties of the material are influenced by its size, leading to altered catalytic performance. Moreover, the composition, shape, and surface structure of nanomaterials can be precisely controlled during their synthesis, providing opportunities to tailor their properties for specific applications.

A wide variety of nanomaterials have been explored for photocatalysis, including metal oxides (such as titanium dioxide, zinc oxide, and iron oxide), metal sulfides (such as cadmium sulfide and copper sulfide), metal nanoparticles (such as gold, silver, and palladium), carbon-based materials (such as graphene and carbon nitride), and many more. These nanomaterials exhibit unique properties and photocatalytic mechanisms, allowing for a broad range of applications.

One of the most well-known nanomaterials used in photocatalysis is titanium dioxide (TiO2), which has been extensively studied due to its excellent photocatalytic properties, stability, and low toxicity. TiO2 nanoparticles have a wide bandgap energy, which allows them to absorb ultraviolet (UV) light, making them suitable for applications in water and air purification. When irradiated with UV light, TiO2 nanoparticles generate electron-hole pairs, which can then participate in various redox reactions, leading to the degradation of organic pollutants, sterilization of bacteria and viruses, and removal of harmful substances from water and air.

Zinc oxide (ZnO) is another popular nanomaterial used in photocatalysis. ZnO exhibits a similar wide bandgap energy as TiO2, allowing it to absorb UV light and generate electron-hole pairs for promoting photocatalytic reactions. ZnO nanoparticles have been used for various applications, such as degradation of organic pollutants, disinfection of water and air, and even in energy conversion devices like solar cells.

Metal nanoparticles, such as gold (Au) and silver (Ag), have also been extensively studied for photocatalytic applications. Metal nanoparticles can exhibit plasmonic properties, where they can strongly absorb and scatter light, leading to localized electromagnetic fields that can enhance photocatalytic activity. Additionally, metal nanoparticles can act as electron sinks, allowing for efficient charge transfer between the photocatalyst and the reactants. Metal nanoparticles have been used in a variety of applications, such as oxidation of organic pollutants, reduction of carbon dioxide, and even in catalytic converters for automotive emissions control.

Carbon-based nanomaterials, such as graphene and carbon nitride, have also shown great promise for photocatalytic applications. Graphene, a two-dimensional material made of a single layer of carbon atoms, exhibits high electrical conductivity and a large surface area, making it suitable for catalytic applications. Graphene-based photocatalysts have been used for degradation of organic pollutants, hydrogen production, and carbon dioxide reduction. Carbon nitride, on the other hand, is a polymeric material that exhibits semiconducting properties and can be synthesized in various forms, such as nanoparticles, nanosheets, and nanorods. Carbon nitride-based photocatalysts have been used for applications such as water splitting, hydrogen production, and degradation of organic pollutants.

Despite the potential advantages of using nanomaterials in photocatalysis, there are also several challenges that must be overcome to realize their full potential. One of the primary challenges is the difficulty in synthesizing nanomaterials with uniform size, shape, and composition. The properties of nanomaterials are strongly influenced by their size and shape, and even slight variations in synthesis conditions can result in significant changes in their properties. Another challenge is the potential toxicity of some nanomaterials, which can limit their use in certain applications. Moreover, the scalability and cost of producing nanomaterials can also be a challenge for widespread adoption of nanomaterials in industrial applications.

In conclusion, nanomaterials photocatalysis is a rapidly growing field of research that holds tremendous promise for addressing various environmental and energy challenges. Nanomaterials offer unique properties that can enhance their photocatalytic activity, making them highly attractive for a wide range of applications. Despite the challenges that must be overcome, the potential benefits of using nanomaterials in photocatalysis are significant, and continued research in this area holds great promise for advancing sustainable technologies.

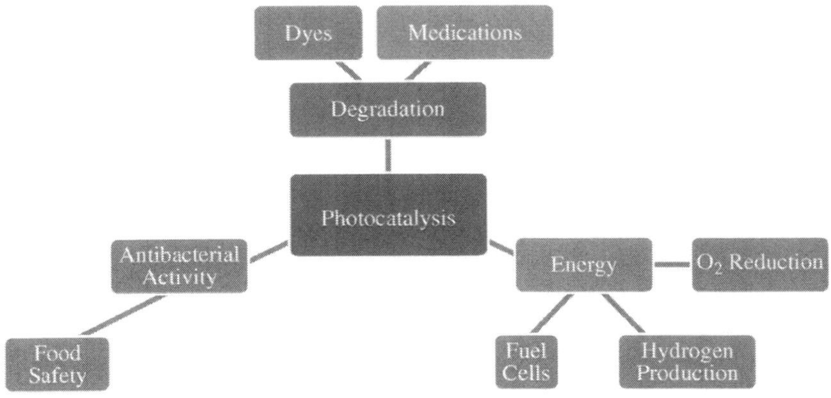

Fig 27. Photocatalysts

10.4 Applications of Nanocatalysis

Nanocatalysis is a rapidly growing field that involves the use of nanoscale materials as catalysts for chemical reactions. The unique properties of nanoscale materials, such as their high surface area, size-dependent electronic and optical properties, and synergistic effects, make them highly effective catalysts for a wide range of chemical reactions. This has led to numerous applications of nanocatalysis in various fields, including energy production, environmental remediation, pharmaceuticals, and materials science.

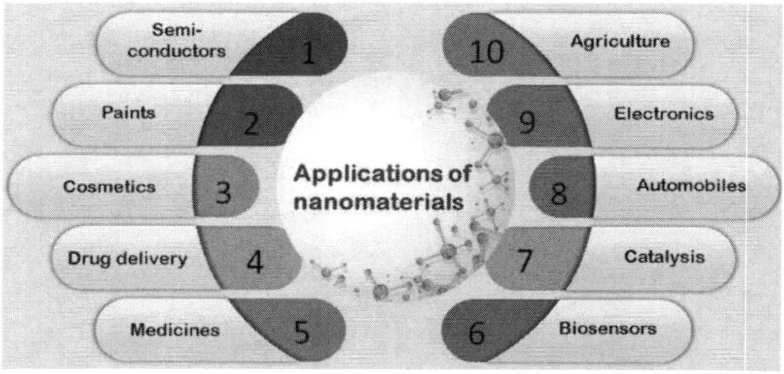

Fig 28. Applications of nanomaterials

Energy production:

One of the most significant applications of nanocatalysis is in the field of energy production. Nanocatalysts are used in the production of fuels, such as hydrogen and biodiesel. For example, platinum nanoparticles are used as catalysts in proton exchange membrane fuel cells to convert hydrogen into electricity. This is because platinum has excellent catalytic activity for the electrochemical oxidation of hydrogen, and its high surface area allows for a greater number of active sites for the reaction to occur.

Additionally, nanocatalysts can be used to convert biomass into biofuels, such as ethanol and biodiesel. This is an attractive alternative to traditional fossil fuels, as biofuels are renewable and produce fewer greenhouse gas emissions. Nanocatalysts have been shown to improve the efficiency of biofuel production and reduce the amount of waste generated during the process.

Environmental remediation:

Nanocatalysts are also used to clean up environmental pollutants, such as heavy metals and organic contaminants, from soil and water. For example, iron oxide nanoparticles are used to remove arsenic from contaminated water. The high surface area and reactivity of the nanoparticles allow them to effectively bind to the arsenic ions, making them easier to remove from the water. Additionally, palladium nanoparticles can be used to break down harmful pollutants, such as trichloroethylene. This is because palladium is a highly effective catalyst for the oxidation of organic compounds, and the use of nanoscale palladium particles allows for greater efficiency and lower cost compared to traditional remediation methods.

Pharmaceuticals:

Nanocatalysts are also used in the synthesis of pharmaceuticals and in drug delivery systems. For example, gold nanoparticles are used as catalysts in the synthesis of anti-cancer drugs. Gold is an excellent catalyst for certain chemical reactions due to its unique electronic and optical properties, and the use of gold nanoparticles allows for greater control over the reaction conditions and product properties. Additionally, silica nanoparticles are used in drug delivery systems to improve drug solubility and bioavailability. The high surface area of the nanoparticles allows them to effectively bind to drugs and protect them from degradation in the body, increasing their effectiveness.

Materials science:

Nanocatalysts are also used in the production of advanced materials, such as carbon nanotubes and graphene. For example, iron nanoparticles can be used as catalysts in the synthesis of carbon nanotubes. This is because the high surface area of the nanoparticles allows for greater control over the growth and properties of the carbon nanotubes, resulting in higher quality and more consistent products. Additionally, nanocatalysts are used in the production of catalyst supports for industrial catalysts, such as those used in the production of chemicals, fertilizers, and plastics.

Chemical manufacturing:

Nanocatalysts are also used in the production of chemicals, such as fertilizers, plastics, and polymers. For example, nickel nanoparticles are used as catalysts in the production of ammonia for fertilizer. This is because nickel has excellent catalytic activity for the Haber-Bosch process, which is used to produce ammonia from nitrogen and hydrogen. Additionally, palladium nanoparticles are used to produce polypropylene, a common plastic. The high surface area and reactivity of the nanoparticles allows for greater efficiency and lower cost compared to traditional methods.

Renewable energy:

In addition to traditional energy production, nanocatalysts are also used in the production of renewable energy sources. For example, nanocatalysts are used in the production of biofuels, such as biodiesel and bioethanol, which are renewable and produce fewer greenhouse gas emissions than traditional fossil fuels. Nanocatalysts are also used in the production of solar cells, which convert sunlight into electricity. For example, titanium dioxide nanoparticles are used as catalysts in dye-sensitized solar cells. The high surface area and reactivity of the nanoparticles allows for greater efficiency in the conversion of sunlight into electricity.

Water treatment:

Nanocatalysts are also used in the treatment of water for drinking and other uses. For example, silver nanoparticles are used as catalysts in the disinfection of water. Silver has antimicrobial properties, and the use of silver nanoparticles allows for greater efficiency in the removal of harmful microorganisms from water. Additionally, nanocatalysts are used in the treatment of wastewater, such as the removal of organic pollutants from industrial wastewater.

Food industry:

Nanocatalysts are also used in the food industry, particularly in the production of food packaging materials. For example, silver nanoparticles are used as catalysts in the production of antimicrobial food packaging materials. The high surface area and reactivity of the nanoparticles allows for greater efficiency in the inhibition of bacterial growth, improving food safety and reducing waste.

CHALLENGES

What are nanocatalysts?

How are nanocatalysts synthesized?

What are the different synthesis methods for nanocatalysts?

How is the size and morphology of nanocatalysts controlled during synthesis?

What are the commonly used materials for nanocatalyst synthesis?

What are the most commonly used characterization techniques for nanocatalysts?

How does the size and morphology of nanocatalysts affect their catalytic activity?

What are the potential applications of nanocatalysts in chemical reactions?

How can nanocatalysts be used in energy production and conversion?

How can nanocatalysts be used in environmental remediation?

How can nanocatalysts be used in the pharmaceutical industry?

What is the role of nanocatalysts in biorefinery processes?

What are the potential challenges of using nanocatalysts in real-world applications?

What are the potential ethical and safety concerns of using nanocatalysts?

What is the future of nanocatalysts in the field of catalysis and beyond?

How do nanocatalysts compare to conventional catalysts in terms of their properties and applications?

How do environmental factors affect the properties and behavior of nanocatalysts?

What are the current trends in research and development of nanocatalysts?

How can nanocatalysts be optimized for specific applications?

What are the key factors that determine the performance of nanocatalysts?

CHAPTER ELEVEN

11.1 Nanomaterials for Energy Applications

Nanomaterials are being extensively explored for various energy applications due to their unique size-dependent properties. They offer enhanced energy efficiency, high performance, and cost-effectiveness compared to traditional materials. Nanomaterials have been applied in different fields such as solar cells, batteries, fuel cells, thermoelectrics, and catalysis. These materials can be designed and engineered to meet specific requirements, leading to efficient energy utilization and storage. The integration of nanomaterials in energy systems can contribute significantly to sustainable energy production and conservation.

11.2 Solar Cells

Nanomaterials have attracted significant attention in the field of solar energy as they offer many advantages over conventional materials. The use of nanomaterials in solar cells has been extensively studied due to their unique electronic, optical, and structural properties.

One application of nanomaterials in solar cells is in the development of next-generation solar cells, such as dye-sensitized solar cells (DSSCs), organic photovoltaics (OPVs), and perovskite solar cells. Nanomaterials such as quantum dots, nanowires, and carbon nanotubes have been used to enhance the efficiency of these solar cells by improving light absorption, charge separation and transport, and reducing recombination losses.

Another application of nanomaterials in solar cells is in the development of transparent conductive electrodes (TCEs). TCEs are a critical component of solar cells, and materials such as indium tin oxide (ITO) are commonly used. However, ITO has limitations such as high cost, scarcity, and brittleness. Nanomaterials such as graphene, carbon nanotubes, and metal nanowires have been explored as alternatives to ITO, offering improved mechanical flexibility, low cost, and high transparency.

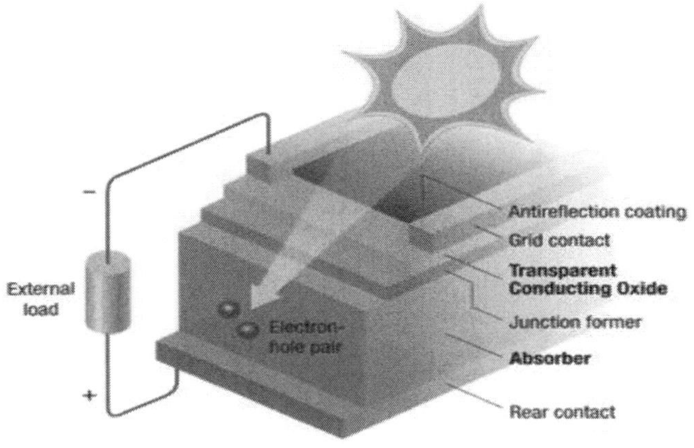

Fig 29. Nanomaterials in solar cells

Nanomaterials have also been used in the development of energy storage devices such as lithium-ion batteries and supercapacitors. Nanomaterials such as graphene, carbon nanotubes, and metal oxides have been used as electrodes in these devices, offering high surface area, improved conductivity, and fast charge/discharge rates.

Overall, the use of nanomaterials in solar cells and energy storage devices has the potential to revolutionize the field of renewable energy, offering improved efficiency, lower costs, and greater scalability.

11.3 Batteries

Nanomaterials have gained significant attention in the field of energy storage and conversion, particularly in batteries. The unique properties of nanomaterials such as high surface area, fast ion transport, and tunable electronic properties make them ideal candidates for battery applications. Some of the ways in which nanomaterials are used in batteries are:

Anode materials: Nanomaterials such as silicon, tin, and germanium have been extensively studied as anode materials due to their high specific capacity. However, these materials suffer from significant volume expansion during charging and discharging, leading to rapid degradation. To overcome this, researchers have developed various nanostructured architectures such as nanowires, nanotubes, and nanoparticles, which can accommodate the volume changes and maintain the structural integrity of the electrode.

Cathode materials: Nanomaterials have also been investigated as cathode materials in batteries. For instance, lithium iron phosphate (LiFePO4) nanoparticles have shown promising results due to their high energy density, long cycle life, and excellent safety. Other examples include nickel cobalt aluminum oxide (NCA) and lithium nickel manganese cobalt oxide (NMC) nanoparticles.

Separator membranes: Nanomaterials such as ceramic nanoparticles and carbon nanotubes have been used to enhance the performance of separator membranes in batteries. These materials can improve the mechanical and thermal stability of the membrane, as well as increase the ionic conductivity.

Electrolytes: Nanomaterials have also been used to improve the performance of electrolytes in batteries. For example, nanoparticles of lithium lanthanum titanate (LLTO) have shown enhanced ionic conductivity and stability at high temperatures, making them suitable for use in solid-state batteries.

Overall, the use of nanomaterials in batteries has the potential to significantly improve the energy density, cycle life, and safety of batteries, leading to their wider adoption in various applications.

11.4 Fuel Cells

Fuel cells are devices that convert chemical energy into electrical energy through the reaction of hydrogen with oxygen or other oxidizing agents. They are gaining significant attention as a potential clean energy source due to their high energy efficiency and low greenhouse gas emissions. However, their commercial viability is limited by high cost and limited durability. Nanomaterials offer potential solutions to these challenges by improving the performance and durability of fuel cells.

Nanomaterials can be used in several parts of fuel cells, including electrodes, electrolytes, and catalysts. One of the main advantages of using nanomaterials is their large surface area, which can enhance electrochemical reactions and increase the reaction rate. Additionally, the unique properties of nanomaterials such as size, shape, and composition can be tailored to improve fuel cell performance.

One of the most important applications of nanomaterials in fuel cells is as catalysts. Catalysts are necessary to enhance the rate of chemical reactions in fuel cells, and nanomaterials have been shown to significantly improve catalyst performance. For example, nanoparticles of platinum (Pt) have been used as a catalyst in proton exchange membrane fuel cells (PEMFCs) due to their high surface area and excellent electrocatalytic activity. However, the high cost and limited durability of Pt catalysts have led researchers to explore alternative nanomaterials such as palladium (Pd), nickel (Ni), and cobalt (Co) nanoparticles. These nanomaterials have demonstrated comparable or better performance than Pt, while also being more abundant and less expensive.

In addition to catalysts, nanomaterials can also be used as electrode materials in fuel cells. Nanomaterials such as carbon nanotubes (CNTs) and graphene have been explored as alternatives to traditional electrode materials due to their high conductivity and large surface area. For example, CNTs have been used as supports for Pt catalysts in PEMFCs, leading to improved catalyst performance and higher power output.

Nanomaterials also have potential applications as electrolytes in fuel cells. Solid oxide fuel cells (SOFCs) use a solid electrolyte to transport ions between the anode and cathode, and nanomaterials such as ceramic oxides have been explored as potential electrolyte materials due to their high ionic conductivity and stability at high temperatures. Additionally, nanomaterials such as carbon-based materials and metal-organic frameworks (MOFs) have been investigated as potential electrolytes in proton-conducting fuel cells due to their high proton conductivity and low cost.

Another potential application of nanomaterials in fuel cells is as gas diffusion layers (GDLs). GDLs are porous materials that allow reactants to reach the electrode while also providing mechanical support for the fuel cell. Nanomaterials such as carbon nanofibers and graphene have been explored as potential GDL materials due to their high surface area and high electrical conductivity. n

Overall, nanomaterials offer significant potential to improve the performance and durability of fuel cells. However, several challenges must be addressed before they can be widely implemented in commercial fuel cells. These include optimizing nanomaterial properties, ensuring their stability and durability under operating conditions, and developing scalable synthesis and manufacturing methods. Nonetheless, the promising results from current research on nanomaterials in fuel cells suggest that they have the potential to significantly impact the development of clean and efficient energy sources.

CHALLENGES

What are nanomaterials for energy applications?

What are the different types of nanomaterials used in energy applications?

How are nanomaterials synthesized for energy applications?

What are the most commonly used characterization techniques for nanomaterials used in energy applications?

What are the potential applications of nanomaterials in energy production and conversion?

How can nanomaterials be used in solar energy harvesting?

How can nanomaterials be used in fuel cells?

How can nanomaterials be used in batteries?

How can nanomaterials be used in energy storage devices?

What are the potential challenges of using nanomaterials in energy applications?

What are the potential ethical and safety concerns of using nanomaterials in energy applications?

What is the future of nanomaterials in the field of energy production and conversion?

How do nanomaterials compare to conventional materials in terms of their properties and applications in energy production and conversion?

CHAPTER TWELVE

12.1 Nanomaterials in Biomedical Applications

Nanomaterials have found numerous applications in various fields of science and technology, including biomedical applications. They have been extensively studied for their potential use in diagnosis, therapy, and drug delivery in the field of medicine. The unique physical and chemical properties of nanomaterials, including their small size, large surface area, and tunable surface chemistry, make them highly attractive for biomedical applications.

12.2 Targeted Drug Delivery using Nanoparticles

One of the major advantages of using nanomaterials in medicine is their ability to penetrate cell membranes and deliver therapeutic agents directly to the target cells. This is particularly important in cancer therapy, where the goal is to selectively target cancer cells while minimizing damage to healthy cells. Nanoparticles can be functionalized with specific targeting ligands such as antibodies, peptides, or aptamers to selectively bind to cancer cells, thereby increasing the efficacy of the therapy. Targeted drug delivery using nanoparticles is a promising approach in the field of drug delivery systems. Nanoparticles offer unique advantages such as high drug loading capacity, sustained drug release, and the ability to target specific cells or tissues. Targeted drug delivery can reduce the side effects of drugs, increase their bioavailability, and improve patient compliance. To achieve targeted drug delivery, nanoparticles can be functionalized with ligands, such as antibodies or peptides, that can recognize and bind to specific receptors on the target cells. The ligands can also be conjugated with therapeutic agents, such as chemotherapy drugs or siRNA, to specifically deliver them to the target cells.

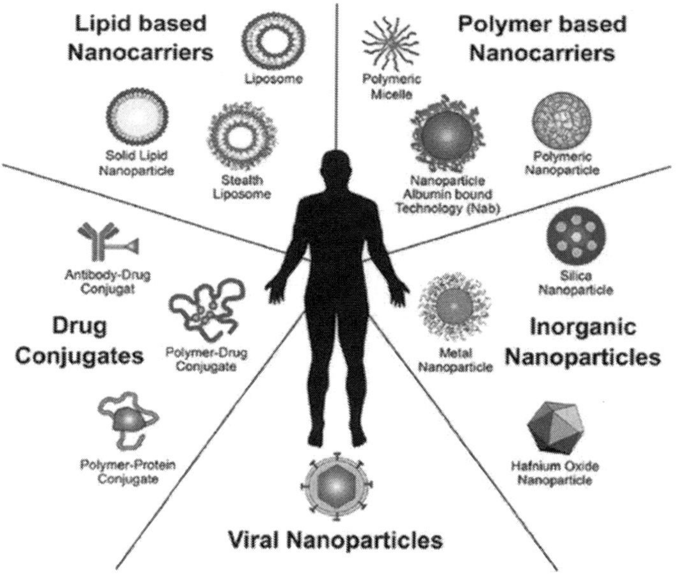

Fig 30. Drug delivery application of nanomaterials

Various types of nanoparticles have been used for targeted drug delivery, such as liposomes, polymeric nanoparticles, dendrimers, and inorganic nanoparticles (e.g., gold nanoparticles). Polymeric nanoparticles can be engineered to release drugs in response to specific stimuli, such as changes in pH or temperature, or the presence of enzymes or other biomolecules. Dendrimers can be designed to have multiple functional groups that can bind to different targets simultaneously, allowing for more specific targeting. Inorganic nanoparticles, such as gold nanoparticles, can be functionalized with peptides that can recognize and bind to cancer cells, and then irradiated with light to induce localized hyperthermia and enhance drug release.

12.3 Imaging and Diagnosis using Nanoparticles

Nanoparticles have been extensively researched and developed for their potential applications in imaging and diagnosis. Due to their small size and unique properties, nanoparticles offer a wide range of benefits over traditional imaging agents.

One of the primary advantages of using nanoparticles for imaging and diagnosis is their ability to specifically target diseased tissues or cells. This is achieved through the use of surface modifications such as ligands, antibodies, or peptides, which allow for targeted delivery and binding to specific receptors on the surface of the diseased cells.

In addition to targeting, nanoparticles can also enhance contrast in imaging techniques such as magnetic resonance imaging (MRI), computed tomography (CT), and optical imaging. This is because nanoparticles can be designed to have high magnetic or optical properties, which can enhance the signal in the imaging technique and improve the detection of the diseased tissue.

Furthermore, nanoparticles can also be used as sensors for disease detection. For example, fluorescent nanoparticles can be designed to respond to certain molecules or changes in the environment, allowing for the detection of specific disease biomarkers or changes in cellular activity.

Nanoparticles have also been explored for their potential in theranostics, which involves combining therapy and diagnosis into a single platform. This is achieved through the use of nanoparticles that can both detect and treat the diseased tissue. For example, nanoparticles can be designed to deliver drugs or genes to the diseased cells while also providing imaging contrast to monitor the therapy's effectiveness.

Overall, the use of nanoparticles in imaging and diagnosis offers several potential benefits, including targeted delivery, enhanced contrast, and the ability to act as sensors for disease detection. Ongoing research in this field holds great promise for the development of new and improved imaging and diagnostic tools for a range of diseases.

12.4 Tissue engineering using nanoparticles

Tissue engineering is an interdisciplinary field that aims to regenerate damaged tissues or organs using biological materials and cells. Nanoparticles, due to their unique properties, have emerged as promising candidates for tissue engineering applications. They can be engineered to interact with cells and tissues at a molecular level and can help improve the mechanical properties and functionality of the engineered tissues.

One of the key applications of nanoparticles in tissue engineering is to enhance the delivery of cells to the target site.

Nanoparticles can be functionalized with specific ligands that can target specific cells or tissues and promote their uptake. This can improve the efficiency of cell delivery and help improve the tissue regeneration process.

Another important application of nanoparticles in tissue engineering is to improve the mechanical properties of the engineered tissues. Nanoparticles can be incorporated into the scaffold materials used for tissue engineering to improve their stiffness, strength, and toughness. This can help improve the mechanical properties of the engineered tissues and make them more suitable for implantation.

Nanoparticles can also be used to provide cues to the cells and promote their differentiation into specific cell types. For example, magnetic nanoparticles can be used to create magnetic fields that can promote the alignment of cells and tissues. Similarly, nanoparticles can be engineered to release specific growth factors or other signaling molecules that can promote the differentiation of cells into specific cell types.

Nanoparticles can also be used to create smart materials that can respond to specific stimuli. For example, pH-responsive nanoparticles can be used to create materials that can change their properties in response to changes in pH. This can be useful for creating materials that can release drugs or other molecules in response to specific physiological conditions.

One of the key challenges in tissue engineering is the development of vascularized tissues. Nanoparticles can be used to create vascularized tissues by promoting the growth of blood vessels. For example, nanoparticles can be functionalized with specific ligands that can promote the migration of endothelial cells and the formation of blood vessels.

In conclusion, the use of nanoparticles in tissue engineering has shown great promise in improving the efficiency and efficacy of tissue regeneration. The unique properties of nanoparticles can be harnessed to improve cell delivery, promote differentiation, improve mechanical properties, and create smart materials that can respond to specific stimuli. Further research is needed to fully realize the potential of nanoparticles in tissue engineering applications.

CHALLENGES

- What are nanomaterials for biomedical applications?
- What are the different types of nanomaterials used in biomedical applications?
- How are nanomaterials synthesized for biomedical applications?
- What are the most commonly used characterization techniques for nanomaterials used in biomedical applications?
- What are the potential applications of nanomaterials in medicine and healthcare?
- How can nanomaterials be used in drug delivery?
- How can nanomaterials be used in cancer therapy?
- How can nanomaterials be used in imaging and diagnostics?
- How can nanomaterials be used in tissue engineering and regenerative medicine?
- What are the potential challenges of using nanomaterials in biomedical applications?
- What are the potential ethical and safety concerns of using nanomaterials in biomedical applications?
- What is the future of nanomaterials in the field of medicine and healthcare?
- How do nanomaterials compare to conventional materials in terms of their properties and applications in medicine and healthcare?

CHAPTER THIRTEEN

13.1 Nanosensors and Nanoelectronics

Nanosensors and nanoelectronics are two important fields in nanotechnology that have been revolutionizing various industries such as healthcare, energy, and electronics. Nanosensors refer to devices that can detect and measure changes in various physical, chemical, and biological parameters at the nanoscale level. Nanoelectronics, on the other hand, deals with the development and use of electronic devices at the nanoscale, which exhibit unique properties that are not observed in conventional electronic devices.

The development of nanosensors and nanoelectronics has been made possible by the unique properties of nanomaterials such as their large surface area, high reactivity, and the ability to tailor their properties by controlling their size, shape, and composition. These properties have enabled the development of highly sensitive and specific sensors and electronic devices that have numerous applications in various fields.

In the field of nanosensors, various types of nanomaterials such as carbon nanotubes, graphene, and quantum dots have been used to develop sensors that can detect changes in temperature, pressure, humidity, gas concentration, and biological molecules such as proteins and DNA. Nanosensors have numerous applications in healthcare, environmental monitoring, and security.

In the field of nanoelectronics, the unique properties of nanomaterials such as high electron mobility, tunable bandgap, and high surface-to-volume ratio have enabled the development of electronic devices such as transistors, diodes, and memories with superior performance compared to their conventional counterparts. For example, carbon nanotube transistors have been shown to exhibit high-speed operation, low power consumption, and excellent thermal stability, making them ideal for use in high-performance electronic devices.

Fig 31. Nanochips

13.2 Synthesis and Characterization Nanosensors

Synthesis of Nanosensors

The synthesis of nanosensors and nanoelectronics involves several techniques, including top-down and bottom-up approaches.

Top-down approaches involve the fabrication of devices by carving or cutting materials down to the desired size. These techniques include lithography, which uses light or electron beams to pattern materials at the nanoscale. Other top-down techniques include etching, grinding, and polishing, which can be used to shape and refine the devices.

Bottom-up approaches involve building devices by assembling individual molecules or atoms. These techniques include self-assembly, where molecules spontaneously organize themselves into a specific structure, and molecular beam epitaxy, where atoms are deposited onto a substrate to build up a structure atom-by-atom.

Regardless of the synthesis approach used, the materials used in nanosensors and nanoelectronics are typically chosen for their unique properties at the nanoscale. For example, materials such as graphene and carbon nanotubes are known for their excellent electrical conductivity, while quantum dots are known for their ability to emit light at specific wavelengths.

The synthesis process for nanosensors and nanoelectronics typically involves several steps, including material synthesis, device design, fabrication, and characterization. Characterization is crucial in understanding the properties and performance of the devices and is typically done using a combination of microscopy, spectroscopy, and electrical measurements.

Overall, the synthesis of nanosensors and nanoelectronics is a complex process that requires careful control of materials, processes, and characterization techniques. However, advancements in these areas hold significant promise for future applications in various industries.

There are various synthesis methods used in the development of nanosensors and nanoelectronics, some of which include:

a. Chemical Vapor Deposition (CVD): This is a widely used technique for growing thin films of nanomaterials on substrates by reacting precursor gases in a high-temperature furnace.
b. Electrochemical Deposition: This method involves depositing metals and other materials onto a substrate using an electrochemical process.
c. Self-Assembly: This method involves the spontaneous organization of molecules or nanoparticles into a desired structure using chemical or physical interactions.
d. Physical Vapor Deposition (PVD): This technique involves depositing a thin film of a material onto a substrate by evaporating it in a vacuum chamber.
e. Atomic Layer Deposition (ALD): This is a precise and controlled deposition technique that allows for the growth of thin films with atomic-level accuracy.
f. Lithography: This involves the patterning of materials at the nanoscale using light or electron beams.
g. Sol-Gel Synthesis: This technique involves the formation of a solid material from a liquid precursor solution by gelation and subsequent drying.
h. Molecular Beam Epitaxy (MBE): This is a technique for growing crystalline films of materials on a substrate by depositing atoms or molecules onto the substrate under ultra-high vacuum conditions.

These are just a few examples of the synthesis methods used in the development of nanosensors and nanoelectronics. The choice of method will depend on the desired properties and application of the device being developed.

Characterization of nanosensors and nanoelectronics

The characterization of nanosensors and nanoelectronics is critical in understanding their properties and performance. The characterization process involves the use of various techniques, including microscopy, spectroscopy, and electrical measurements.

a. Microscopy: Microscopy techniques such as scanning electron microscopy (SEM), transmission electron microscopy (TEM), and atomic force microscopy (AFM) can be used to visualize the morphology and structure of the devices at the nanoscale.
b. Spectroscopy: Spectroscopy techniques such as Raman spectroscopy, infrared spectroscopy (IR), and X-ray photoelectron spectroscopy (XPS) can be used to analyze the chemical composition and surface properties of the devices. These techniques can provide information on the functional groups present in the device and their bonding states.
c. Electrical Measurements: Electrical measurements such as current-voltage (I-V) curves, impedance spectroscopy, and conductance measurements can be used to evaluate the electrical properties of the devices. These measurements can provide information on the conductivity, resistance, and capacitance of the device.
d. Thermal Analysis: Thermal analysis techniques such as thermogravimetric analysis (TGA) and differential scanning calorimetry (DSC) can be used to evaluate the thermal properties of the devices. These techniques can

provide information on the thermal stability, melting point, and thermal conductivity of the device.

e. Surface Analysis: Surface analysis techniques such as ellipsometry and contact angle measurements can be used to analyze the surface properties of the devices. These techniques can provide information on the surface energy, roughness, and hydrophobicity of the device.

Overall, the characterization of nanosensors and nanoelectronics involves a combination of techniques to provide a comprehensive understanding of the device's properties and performance. The characterization process is critical in the development of these technologies, as it enables researchers to optimize device performance and improve their functionality for a range of applications.

13.3 Applications of Nanosensors and Nanoelectronics

Nanosensors and nanoelectronics have a wide range of potential applications in various fields, some of which include:

a. Biomedical Applications: Nanosensors and nanoelectronics can be used in the detection and diagnosis of diseases. For example, nanosensors can detect specific biomarkers in biological fluids, while nanoelectronics can be used for implantable medical devices such as pacemakers and glucose monitors.
b. Environmental Monitoring: Nanosensors can be used to detect pollutants and monitor environmental parameters such as temperature, humidity, and air quality.
c. Energy Applications: Nanoelectronics can be used in the development of energy-efficient devices and renewable energy technologies such as solar cells.

d. Aerospace and Defense: Nanosensors and nanoelectronics can be used in the development of lightweight and durable materials for aerospace applications, as well as in the detection of explosives and chemical warfare agents.
e. Consumer Electronics: Nanoelectronics can be used in the development of smaller, faster, and more energy-efficient electronic devices such as smartphones, laptops, and tablets.
f. Food and Agriculture: Nanosensors can be used in the detection of food pathogens and spoilage, as well as in the development of precision agriculture technologies for crop monitoring and management.

Overall, the potential applications of nanosensors and nanoelectronics are vast and varied, with the potential to transform numerous industries and improve our quality of life.

CHALLENGES

- What is a nanosensor?
- How are nanosensors used in various fields such as medicine, environment, and industry?
- What are the different types of nanosensors?
- How are nanosensors fabricated?
- What are some characterization techniques used for nanosensors?
- What are the advantages of using nanosensors over traditional sensors?
- What are the limitations of nanosensors?
- What are some potential applications of nanosensors in medicine?
- How can nanosensors be used in environmental monitoring?
- What are the challenges in developing nanosensors for industrial applications?
- What is a nano-electronic device?
- What are the different types of nano-electronic devices?
- How are nano-electronic devices fabricated?
- What are some characterization techniques used for nano-electronic devices?
- What are the advantages of using nano-electronic devices over traditional electronic devices?
- What are the limitations of nano-electronic devices?
- What are some potential applications of nano-electronics?
- What is the future of nanosensors and nano-electronics?

- How can nanotechnology be used to enhance the performance of electronic devices?
- What are the ethical and safety concerns surrounding the use of nanosensors and nano-electronics?

CHAPTER FOURTEEN

14.1 Nanomagnetism

Nanomagnetism is the study of magnetic phenomena at the nanoscale, typically in materials with dimensions ranging from a few nanometers to several hundred nanometers. In these materials, the properties of the magnetic structure are strongly influenced by quantum mechanical effects, which can result in unique magnetic behaviors that are not observed in bulk materials.

The fundamental principles of nanomagnetism are rooted in the behavior of electron spins, which are the quantum mechanical property that gives rise to magnetism. In nanomagnetic materials, the behavior of electron spins is strongly influenced by surface effects, quantum confinement, and magnetic interactions between neighboring atoms or nanoparticles.

Overall, nanomagnetism is a rapidly growing field that is advancing our understanding of the fundamental properties of magnetic materials and opening up new avenues for technological innovation in a wide range of applications.

14.2 Synthesis and Characterization of Nanomagnetic Materials

Synthesis of nanomagnetic materials

The synthesis of nanomagnetic materials involves the preparation of nanoparticles or nanocrystals of magnetic materials with dimensions in the range of 1-100 nanometers. These materials have unique properties due to their small size, including high surface area, enhanced magnetic properties, and potential applications in various fields, such as data storage, biomedicine, and energy conversion.

There are several methods for synthesizing nanomagnetic materials, including chemical precipitation, sol-gel, hydrothermal, and co-precipitation methods. One popular approach is the co-precipitation method, which involves the simultaneous precipitation of metal ions in the presence of a reducing agent and surfactant to form nanoparticles.

There are several methods for synthesizing nanomagnetic materials, some of which are listed below:

a. Co-precipitation method: involves the simultaneous precipitation of metal ions in the presence of a reducing agent and surfactant to form nanoparticles.
b. Sol-gel method: involves the hydrolysis and polycondensation of metal alkoxides in a solvent to form a gel, followed by drying and calcination to produce nanocrystals.
c. Hydrothermal method: involves the reaction of metal precursors in a high-pressure and high-temperature

environment, typically using an autoclave, to produce nanocrystals.

d. Ball milling: involves the mechanical milling of bulk magnetic materials to produce nanocrystals.
e. Electrodeposition: involves the electrochemical deposition of metal ions onto a substrate to form a thin film or nanocrystals.
f. Microwave synthesis: involves the use of microwave radiation to heat the reaction mixture and produce nanocrystals.
g. Spray pyrolysis: involves the atomization of a precursor solution into a high-temperature furnace to produce nanocrystals.
h. Thermal decomposition: involves the decomposition of metal organic precursors under controlled conditions to produce nanocrystals.
i. Plasma arc synthesis: involves the use of a plasma arc to vaporize a metal target and deposit the vapor onto a substrate to produce nanocrystals.
j. Chemical vapor deposition: involves the deposition of metal ions onto a substrate using a chemical vapor deposition process to produce nanocrystals.

Characterization of nanomagnetic materials

Characterization of nanomagnetic materials is an essential step in understanding their properties and potential applications. The following are some of the commonly used techniques for the characterization of nanomagnetic materials:

a. X-ray diffraction (XRD): XRD is used to determine the crystal structure and phase purity of the nanomagnetic materials. This technique provides information about the size, shape, and orientation of the crystals.
b. Transmission electron microscopy (TEM): TEM is used to image the nanomagnetic materials at high resolution and provides information about the size, shape, and

morphology of the nanoparticles. TEM can also be used to determine the crystal structure and lattice parameters of the nanomagnetic materials.

c. Scanning electron microscopy (SEM): SEM is used to image the surface of the nanomagnetic materials and provides information about the size, shape, and morphology of the nanoparticles.
d. Vibrating sample magnetometry (VSM): VSM is used to measure the magnetic properties of the nanomagnetic materials, such as magnetization, coercivity, and remanence.
e. Mössbauer spectroscopy: Mössbauer spectroscopy is used to study the electronic and magnetic properties of the nanomagnetic materials. It provides information about the magnetic moment, hyperfine splitting, and spin relaxation time of the nanoparticles.
f. Atomic force microscopy (AFM): AFM is used to image the surface of the nanomagnetic materials and provides information about the size, shape, and surface roughness of the nanoparticles.
g. Dynamic light scattering (DLS): DLS is used to measure the size distribution of the nanoparticles in a suspension. It provides information about the average size and size distribution of the nanoparticles.

Overall, the characterization techniques listed above, when used together, can provide a comprehensive understanding of the properties of nanomagnetic materials and their potential applications.

14.3 Chemistry of Nanomagnetic Materials

Nanomagnetic materials are typically composed of magnetic metals or metal oxides, such as iron, cobalt, nickel, and their alloys. The chemical composition of nanomagnetic materials plays a crucial role in determining their magnetic properties and potential applications. The following are some of the key aspects of the chemistry of nanomagnetic materials:

Crystal structure: The crystal structure of nanomagnetic materials determines their magnetic properties, such as magnetization and anisotropy. For example, the crystal structure of iron oxide nanoparticles can vary from magnetite (Fe_3O_4) to maghemite (γ-Fe_2O_3), which influences their magnetic properties.

Particle size: The size of nanoparticles determines their magnetic behavior. As the size of the nanoparticles decreases, their magnetic moments become more disordered, and their magnetic properties change.

Surface chemistry: The surface chemistry of nanomagnetic materials plays a significant role in their applications, such as drug delivery and biosensors. Surface modification can alter the surface charge and hydrophobicity of nanoparticles, which affects their interactions with biological molecules.

Magnetic anisotropy: The magnetic anisotropy of nanomagnetic materials arises from their crystal structure and shape. Anisotropic nanoparticles have a preferred orientation of their magnetic moment, which determines their magnetic properties.

Magnetization: The magnetization of nanomagnetic materials is influenced by several factors, including crystal structure, particle size, and temperature. For example, magnetization decreases with increasing temperature and can be used to tune the magnetic properties of nanoparticles.

Overall, the chemistry of nanomagnetic materials is a complex field that involves the study of various factors that influence their magnetic properties and potential applications. Understanding the chemistry of nanomagnetic materials is essential for developing new materials with desired properties and optimizing their applications.

14.4 Applications of Nanomagnetic Materials

Nanomagnetic materials have a wide range of potential applications in various fields due to their unique magnetic properties and small size. Some of the most promising applications of nanomagnetic materials are:

a. Biomedical applications: Nanomagnetic materials can be used in biomedical applications such as targeted drug delivery, magnetic hyperthermia, and magnetic resonance imaging (MRI). Magnetic nanoparticles can be functionalized with biomolecules to target specific cells or tissues for drug delivery or to act as contrast agents for MRI.
b. Data storage: Nanomagnetic materials have applications in data storage, such as in hard disk drives, where the magnetic properties of the nanoparticles are used to store and retrieve data.
c. Magnetic sensors: Nanomagnetic materials can be used in magnetic sensors to detect magnetic fields, such as in magnetic field sensors for navigation, security systems, and medical devices.
d. Environmental applications: Nanomagnetic materials can be used for environmental applications such as water treatment and remediation. Magnetic nanoparticles can be functionalized with chemicals that can bind to pollutants, making them easier to remove from water.
e. Energy applications: Nanomagnetic materials can be used in energy applications such as energy storage and

conversion. Magnetic nanoparticles can be used as catalysts for fuel cells and can also be used to remove impurities from oil and gas.

Overall, the unique magnetic properties of nanomagnetic materials make them attractive for a wide range of applications, from biomedical applications to data storage and environmental remediation. As research in this field continues, new applications for nanomagnetic materials are likely to emerge.

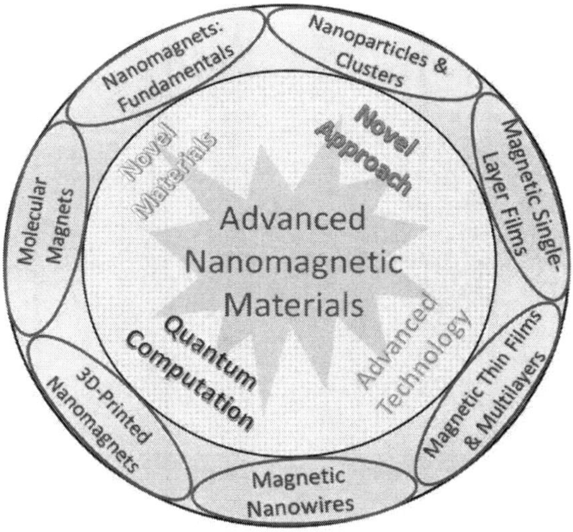

Fig 32. Nanomagnetism

CHALLENGES

- What are nanomagnetic materials?
- What are the different types of nanomagnetic materials?
- How are nanomagnetic materials synthesized?
- What are some characterization techniques used for nanomagnetic materials?
- How do the properties of nanomagnetic materials differ from those of bulk materials?
- What are the advantages of using nanomagnetic materials over traditional magnetic materials?
- What are the limitations of nanomagnetic materials?
- What are some potential applications of nanomagnetic materials?
- How can nanomagnetic materials be used in data storage?
- What are the challenges in developing nanomagnetic materials for biomedical applications?
- What is magnetic resonance imaging (MRI) and how are nanomagnetic materials used in it?
- How can nanomagnetic materials be used in energy applications?
- How can nanomagnetic materials be used in sensing applications?
- How can nanomagnetic materials be used in environmental remediation?
- What are some potential risks associated with the use of nanomagnetic materials?
- How can the toxicity of nanomagnetic materials be assessed?
- What are the ethical considerations surrounding the use of nanomagnetic materials?

- What is the future of nanomagnetic materials?
- How can advances in synthesis and characterization techniques improve the performance of nanomagnetic materials?
- How can nanomagnetic materials be integrated with other nanomaterials to create new functionalities?

CHAPTER FIFTEEN

15.1 Nanoemulsion

Nanoemulsions are a type of emulsion that consists of small droplets of one liquid dispersed in another liquid, where the droplet size is typically in the nanometer range. Nanoemulsions have received increasing attention in recent years due to their unique properties and potential applications in various fields, including food, pharmaceuticals, cosmetics, and biotechnology.

Nanoemulsions are stabilized by the use of surfactants, which lower the interfacial tension between the two liquids and prevent the droplets from coalescing. The small droplet size of nanoemulsions provides several advantages over conventional emulsions, such as improved stability, increased surface area, and enhanced bioavailability.

Nanoemulsions can be prepared using various methods, including high-pressure homogenization, ultrasonication, microfluidization, and phase inversion temperature. The choice of preparation method can affect the droplet size, distribution, and stability of the nanoemulsion.

The small droplet size of nanoemulsions also makes them useful for encapsulating and delivering bioactive compounds, such as drugs, vitamins, and nutraceuticals. Nanoemulsions can protect the bioactive compounds from degradation and enhance their absorption and efficacy.

Overall, nanoemulsions are a promising class of emulsions with unique properties and potential applications in various fields. As research in this area continues, new and exciting applications for nanoemulsions are likely to emerge.

15.2 Synthesis and characterization of nanoemulsions

Synthesis of nanoemulsions

The synthesis of nanoemulsions typically involves the use of high-energy emulsification techniques, such as ultrasonication, high-pressure homogenization, and microfluidization. These techniques can produce nanoemulsions with a narrow size distribution and high stability.

There are several methods for synthesizing nanoemulsions, some of which are listed below:

a. High-pressure homogenization: This method involves using high pressure to force the emulsion through a small orifice, resulting in the reduction of droplet size to the nanometer range.
b. Ultrasonication: This method involves using high-frequency sound waves to generate cavitation, which breaks up the larger droplets into smaller ones.

c. Microfluidization: This method involves using high-pressure pumps to force the emulsion through narrow channels, resulting in the reduction of droplet size.
d. Phase inversion temperature (PIT) method: This method involves changing the temperature to induce a phase transition, resulting in the formation of nanoemulsions.
e. Spontaneous emulsification: This method involves using a surfactant to form an emulsion without the need for external energy input.
f. Membrane emulsification: This method involves using a membrane to form droplets of the dispersed phase, resulting in the formation of nanoemulsions.

Each of these methods has its advantages and disadvantages, and the choice of method depends on factors such as the desired droplet size, stability, and scalability of the process.

Characterization of nanoemulsions

Characterization of nanoemulsions is crucial to ensure their quality, stability, and suitability for various applications. Some common characterization techniques for nanoemulsions are:

a. Particle size analysis: This technique determines the size distribution of the emulsion droplets, typically using dynamic light scattering (DLS) or laser diffraction methods.

b. Zeta potential analysis: This technique measures the electrostatic potential at the surface of the emulsion droplets, which is important for understanding their stability and interactions with other materials.

c. Transmission electron microscopy (TEM): This technique allows visualization of the size, shape, and morphology of individual emulsion droplets, which can provide insights into their stability and behavior.

d. Fourier transform infrared spectroscopy (FTIR): This technique is used to analyze the chemical composition of the emulsion, including the surfactants and other components.

e. Rheological analysis: This technique measures the viscosity and flow behavior of the emulsion, which is important for understanding its stability and potential applications.

f. Stability studies: These studies involve monitoring the emulsion over time to assess its stability and resistance to factors such as temperature changes, pH changes, and mechanical stress.

Overall, a combination of these techniques can provide a comprehensive understanding of the physical, chemical, and functional properties of nanoemulsions.

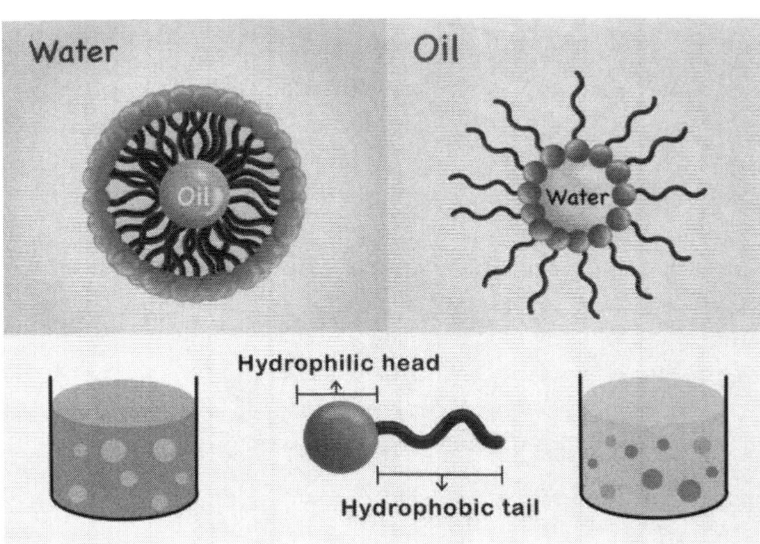

Fig 33. Nanoemulsion

15.3 Chemistry of nanoemulsions

Nanoemulsions are a type of emulsion that consists of small droplets of one liquid dispersed in another liquid. The droplet size of nanoemulsions is typically in the range of 20 to 200 nanometers. Compared to traditional emulsions, nanoemulsions have several advantages, including increased stability, increased surface area, and enhanced bioavailability. These properties make them useful for a wide range of applications, including drug delivery, food processing, and cosmetics.

The chemistry of nanoemulsions involves understanding the interactions between the different components that make up the emulsion, including the dispersed phase, continuous phase, and surfactants. The choice of each component and the conditions under which they are used can have a significant impact on the size, stability, and other properties of the emulsion droplets.

The dispersed phase in a nanoemulsion can be a wide range of materials, including oils, water-insoluble drugs, and other hydrophobic compounds. The choice of dispersed phase can affect the size and stability of the droplets, as well as the compatibility of the emulsion with different applications.

The continuous phase in a nanoemulsion is typically water, but it can also be other polar or nonpolar solvents. The choice of continuous phase can affect the stability, viscosity, and other properties of the emulsion. For example, the viscosity of the continuous phase can affect the ability of the surfactants to adsorb to the interface between the two phases.

Surfactants are key components in the chemistry of nanoemulsions. They are molecules that have both hydrophilic (water-loving) and hydrophobic (water-hating) regions, allowing them to adsorb to the surface of the emulsion droplets and stabilize them by reducing interfacial tension. The choice of surfactant and its concentration can have a significant impact on the stability, size distribution, and other properties of the emulsion droplets.

The type of surfactant used can also affect the mechanism of emulsion formation. For example, some surfactants may form an adsorption layer on the droplet surface, while others may form a micelle structure around the droplet.

The choice of emulsification method can also affect the chemistry of nanoemulsions. High-energy methods, such as ultrasonication, high-pressure homogenization, and microfluidization, are commonly used to produce nanoemulsions. These methods can generate high shear forces that break up larger droplets into smaller ones. However, the shear forces can also cause damage to the surfactant layer, leading to destabilization of the emulsion droplets.

The size, stability, and other properties of nanoemulsions can be characterized using a variety of techniques. Particle size analysis, such as dynamic light scattering (DLS) and laser diffraction, is commonly used to determine the size distribution of the emulsion droplets. Zeta potential analysis can provide information on the stability of the emulsion by measuring the electrostatic potential at the surface of the droplets. Transmission electron microscopy (TEM) can be used to visualize the size and morphology of the individual droplets. Rheological analysis can provide information on the viscosity and flow behavior of the emulsion.

The chemistry of nanoemulsions can also be affected by environmental factors, such as temperature, pH, and ionic strength. These factors can affect the stability of the surfactant layer and the interactions between the dispersed phase and the continuous phase.

Overall, the chemistry of nanoemulsions involves understanding the interactions between the different components that make up the emulsion, and how they can be optimized to achieve desired properties and performance in different applications. The choice of dispersed phase, continuous phase, surfactant, and emulsification method

15.4 Applications of nanoemulsions

Nanoemulsions have a wide range of applications across several industries due to their unique properties, such as increased stability, increased surface area, and enhanced bioavailability. Some of the applications of nanoemulsions are as follows:

Pharmaceutical Industry: Nanoemulsions are widely used in drug delivery systems due to their ability to improve the solubility, bioavailability, and stability of drugs. They are used in both oral and topical formulations for a wide range of drugs, including anticancer drugs, anti-inflammatory drugs, and antibiotics.

Food Industry: Nanoemulsions are used in the food industry to improve the flavor, texture, and stability of food products. They are used in the production of beverages, dairy products, baked goods, and condiments. Nanoemulsions can also be used as carriers for functional ingredients, such as vitamins and antioxidants.

Cosmetics Industry: Nanoemulsions are used in the cosmetics industry to improve the delivery of active ingredients to the skin. They are used in the production of creams, lotions, and sunscreens. Nanoemulsions can improve the penetration of active ingredients into the skin, leading to enhanced efficacy.

Agriculture Industry: Nanoemulsions can be used as a delivery system for pesticides, herbicides, and fertilizers in the agriculture industry. They can improve the effectiveness of these products by enhancing their solubility and reducing their environmental impact.

Oil and Gas Industry: Nanoemulsions can be used in the oil and gas industry to enhance oil recovery. They can be injected into reservoirs to displace trapped oil, leading to increased recovery rates.

Textile Industry: Nanoemulsions can be used in the textile industry to improve the dyeing process. They can improve the solubility and penetration of dyes, leading to improved color uniformity and reduced water usage.

Personal Care Industry: Nanoemulsions can be used in the production of personal care products, such as deodorants, hair care products, and shaving creams. They can improve the delivery of active ingredients to the skin and hair, leading to enhanced efficacy.

In conclusion, the unique properties of nanoemulsions make them useful for a wide range of applications across several industries. As technology continues to advance, we can expect to see even more applications of nanoemulsions in the future.

CHAPTER SIXTEEN

16.1 Nanomaterials for Environmental Applications

Nanomaterials are materials that have at least one dimension in the nanoscale range (1-100 nm). Due to their unique physicochemical properties, such as large surface area to volume ratio, high reactivity, and quantum confinement effects, nanomaterials have shown significant potential in various environmental applications.

One of the most promising applications of nanomaterials is in environmental remediation. Nanomaterials can be used to remove pollutants from water, air, and soil. For example, nanoscale zero-valent iron (nZVI) can be used to degrade organic pollutants, such as chlorinated solvents and pesticides, in contaminated groundwater. Carbon nanotubes (CNTs) can be used to remove heavy metals, such as lead and mercury, from wastewater.

Nanomaterials can also be used for sensing and monitoring of environmental contaminants. They can detect pollutants at low concentrations and in real-time. For example, gold nanoparticles can be used as colorimetric sensors for the detection of heavy metal ions in water. Quantum dots can be used for the detection of pesticides in food samples.

Another promising application of nanomaterials is in renewable energy. Nanomaterials can be used to improve the efficiency of solar cells, fuel cells, and batteries. For example, nanoparticles of titanium dioxide (TiO_2) can be used as a photoanode in dye-sensitized solar cells, which convert sunlight into electricity.

However, the use of nanomaterials in environmental applications also raises concerns about their potential environmental and health risks. The small size and high reactivity of nanomaterials may result in unintended environmental impacts, such as bioaccumulation and toxicity. Therefore, it is important to consider the potential risks associated with the use of nanomaterials in environmental applications and to develop strategies for their safe handling and disposal.

Overall, nanomaterials have shown significant potential in various environmental applications, but their potential risks must also be carefully considered. Ongoing research in this field is needed to further understand the behavior and potential impacts of nanomaterials in the environment.

16.2 Water Purification

Water pollution is a major environmental concern worldwide, with many contaminants present in water sources. Nanomaterials have emerged as promising materials for water purification due to their unique properties such as large surface area to volume ratio, high reactivity, and tunable properties. In this article, we will discuss the various types of nanomaterials and their applications in water purification.

Types of Nanomaterials for Water Purification:

Metal and Metal Oxide Nanoparticles: Metal and metal oxide nanoparticles, such as iron oxide, silver, and copper, have shown significant potential for water purification. These nanoparticles can remove various contaminants, such as heavy metals, organic pollutants, and pathogens, from water sources. For example, silver nanoparticles have been shown to be effective in inhibiting the growth of bacteria and viruses.

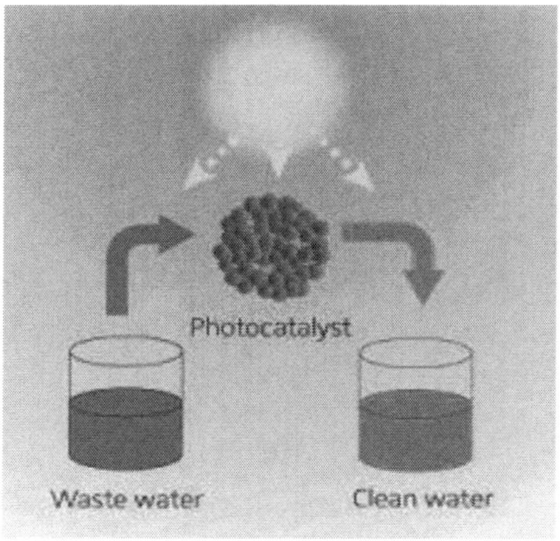

Fig 34. Water purification usung nano

Carbon-Based Nanomaterials: Carbon-based nanomaterials, such as graphene and carbon nanotubes (CNTs), have also shown promise for water purification. These materials have high surface area, mechanical strength, and excellent adsorption properties, making them useful for removing various contaminants from water sources. For example, CNTs have been shown to be effective in removing heavy metals and organic pollutants from water.

Nanocellulose: Nanocellulose is a biodegradable nanomaterial that can be derived from plant-based sources. It has been shown to be effective in removing contaminants such as heavy metals, dyes, and bacteria from water sources. Nanocellulose has the added advantage of being biocompatible and environmentally friendly. 08068649043

Metal-Organic Frameworks (MOFs): MOFs are porous materials composed of metal ions and organic ligands. They have a high surface area and tunable pore size, making them useful for removing various contaminants from water sources. For example, MOFs have been shown to be effective in removing heavy metals, organic pollutants, and dyes from water.

Applications of Nanomaterials in Water Purification:

a. Removal of Heavy Metals: Heavy metals are a major source of water pollution and can have adverse effects on human health. Nanomaterials, such as metal and metal oxide nanoparticles, CNTs, and MOFs, have shown promising results for the removal of heavy metals from water sources. These materials can adsorb heavy metals onto their surface, leading to the removal of contaminants from water sources.
b. Removal of Organic Pollutants: Organic pollutants, such as pesticides and industrial chemicals, are another major source of water pollution. Nanomaterials, such as carbon-based nanomaterials and MOFs, have shown promising results for the removal of organic pollutants from water sources. These materials can adsorb organic pollutants onto their surface, leading to the removal of contaminants from water sources.
c. Removal of Pathogens: Pathogens, such as bacteria and viruses, can cause waterborne diseases and are a major source of water pollution. Nanomaterials, such as silver nanoparticles and nanocellulose, have shown promising

results for the removal of pathogens from water sources. These materials can inhibit the growth of bacteria and viruses, leading to the removal of contaminants from water sources.

d. Desalination: Desalination is the process of removing salt and other minerals from seawater to make it suitable for drinking and irrigation purposes. Nanomaterials, such as graphene and MOFs, have shown promising results for the desalination of seawater. These materials can remove salt and other minerals from seawater by using their high surface area and tunable pore size.

Nanomaterials have emerged as promising materials for water purification due to their unique properties and tunable properties. Metal and metal oxide nanoparticles, carbon-based nanomaterials, nanocellulose

16.3 Air Purification

Air pollution is a major environmental concern worldwide, with many pollutants present in the air we breathe. Nanomaterials have emerged as promising materials for air purification due to their unique properties such as large surface area to volume ratio, high reactivity, and tunable properties. In this article, we will discuss the various types of nanomaterials and their applications in air purification.

Types of Nanomaterials for Air Purification:

Metal and Metal Oxide Nanoparticles: Metal and metal oxide nanoparticles, such as titanium dioxide and zinc oxide, have shown significant potential for air purification. These nanoparticles can remove various pollutants, such as volatile organic compounds (VOCs) and nitrogen oxides (NOx), from the air. For example, titanium dioxide nanoparticles have been shown to be effective in removing pollutants from indoor air.

Carbon-Based Nanomaterials: Carbon-based nanomaterials, such as activated carbon and carbon nanotubes (CNTs), have also shown promise for air purification. These materials have high surface area, mechanical strength, and excellent adsorption properties, making them useful for removing various pollutants from the air. For example, activated carbon has been widely used in air filters to remove pollutants such as VOCs and particulate matter.

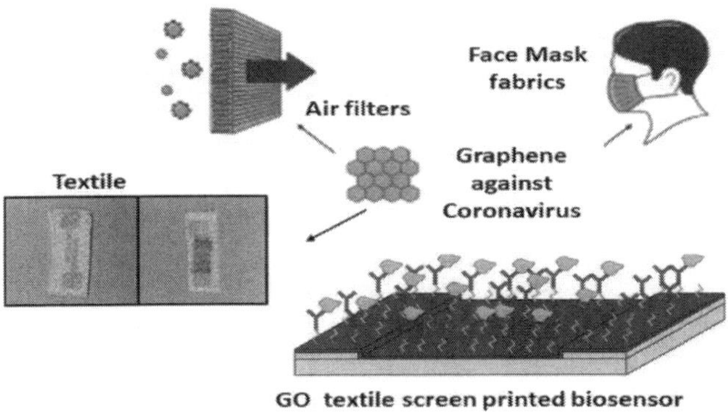

Fig 35. Nanomaterials for air purification

Nanoporous Materials: Nanoporous materials, such as zeolites and metal-organic frameworks (MOFs), have a high surface area and tunable pore size, making them useful for removing various pollutants from the air. For example, zeolites have been shown to be effective in removing VOCs from the air.

Applications of Nanomaterials in Air Purification:

a. Indoor Air Purification: Indoor air pollution is a major health concern, with many pollutants present in indoor environments. Nanomaterials, such as metal and metal oxide nanoparticles, carbon-based nanomaterials, and nanoporous materials, have shown promise for indoor air purification. These materials can remove various pollutants, such as VOCs, NOx, and particulate matter, from indoor air.

b. Outdoor Air Purification: Outdoor air pollution is a major environmental concern, with many pollutants present in the air we breathe. Nanomaterials, such as metal and metal oxide nanoparticles and nanoporous materials, have shown promise for outdoor air purification. For example, titanium dioxide nanoparticles have been used in pavement to reduce air pollution by removing pollutants from the air.
c. Filtration Systems: Nanomaterials have also been used in air filtration systems to remove pollutants from the air. For example, carbon-based nanomaterials, such as activated carbon and CNTs, have been widely used in air filters to remove pollutants such as VOCs and particulate matter.
d. Catalytic Converters: Catalytic converters are devices that convert pollutants in vehicle exhaust into less harmful substances. Nanomaterials, such as metal and metal oxide nanoparticles, have shown promise for use in catalytic converters to improve their efficiency and reduce their size and cost.
e. Nanomaterials have emerged as promising materials for air purification due to their unique properties and tunable properties. Metal and metal oxide nanoparticles, carbon-based nanomaterials, and nanoporous materials have shown significant potential for removing pollutants from both indoor and outdoor air environments. Further research and development are needed to fully understand the potential of nanomaterials for air purification and to ensure their safety and environmental impact.

CHALLENGES

- What are nanomaterials for water and air purification?
- How do nanomaterials help in the purification of water and air?
- What are the different types of nanomaterials used for water and air purification?
- How are nanomaterials synthesized for water and air purification?
- What are some characterization techniques used for nanomaterials for water and air purification?
- What are the advantages of using nanomaterials for water and air purification over traditional purification methods?
- What are the limitations of nanomaterials for water and air purification?
- How can nanomaterials be used to remove specific contaminants from water and air?
- What are the challenges in developing nanomaterials for water and air purification?
- How can nanomaterials be used in desalination processes?
- How can nanomaterials be used in wastewater treatment?
- What are the potential risks associated with the use of nanomaterials for water and air purification?
- How can the toxicity of nanomaterials be assessed?
- What are the ethical considerations surrounding the use of nanomaterials for water and air purification?
- What are some potential applications of nanomaterials for air purification?
- What are some potential applications of nanomaterials for water purification?

- How can advances in synthesis and characterization techniques improve the performance of nanomaterials for water and air purification?
- How can nanomaterials be integrated with other nanomaterials or traditional purification methods to create new functionalities?
- What is the future of nanomaterials for water and air purification?
- How can nanomaterials for water and air purification be made more accessible and affordable for developing countries?

CHAPTER SEVENTEEN

17.1 Nanotoxicology

Nanotoxicology is a field of study that investigates the potential adverse effects of nanomaterials on living organisms, including humans and the environment. With the rapid development of nanotechnology, the use of nanomaterials in various applications has increased dramatically, including in consumer products, medicine, and industry. As a result, there is a growing need to understand the potential risks associated with the use of these materials.

Nanomaterials are unique in that their small size (typically less than 100 nanometers) and high surface area to volume ratio can result in unique physicochemical properties compared to their bulk counterparts. These properties can potentially lead to increased toxicity, as nanomaterials may be more likely to interact with biological systems in unintended ways. For example, some studies have shown that certain types of nanoparticles can cause inflammation, oxidative stress, and damage to DNA and cells in vitro and in vivo.

Nanotoxicology seeks to investigate the potential hazards and risks associated with exposure to nanomaterials, including their toxicity, mechanisms of action, and potential environmental impacts. It involves the use of various techniques and methods, such as in vitro and in vivo toxicity testing, physicochemical characterization, and risk assessment, to Nanotoxicology is a field of study that investigates the potential adverse effects of nanomaterials on living organisms, including humans and the environment. With the rapid development of nanotechnology, the use of nanomaterials in various applications has increased dramatically, including in consumer products, medicine, and industry. As a result, there is a growing need to understand the potential risks associated with the use of these materials.

Nanomaterials are unique in that their small size (typically less than 100 nanometers) and high surface area to volume ratio can result in unique physicochemical properties compared to their bulk counterparts. These properties can potentially lead to increased toxicity, as nanomaterials may be more likely to interact with biological systems in unintended ways. For example, some studies have shown that certain types of nanoparticles can cause inflammation, oxidative stress, and damage to DNA and cells in vitro and in vivo.

Nanotoxicology seeks to investigate the potential hazards and risks associated with exposure to nanomaterials, including their toxicity, mechanisms of action, and potential environmental impacts. It involves the use of various techniques and methods, such as in vitro and in vivo toxicity testing, physicochemical characterization, and risk assessment, to better understand the behavior and effects of nanomaterials.

One major challenge in nanotoxicology is the lack of standardized methods and protocols for testing the toxicity of nanomaterials. Unlike traditional chemicals, nanomaterials often exhibit complex physicochemical properties that can influence their behavior and toxicity. As a result, standardized methods and guidelines for nanotoxicity testing are still under development.

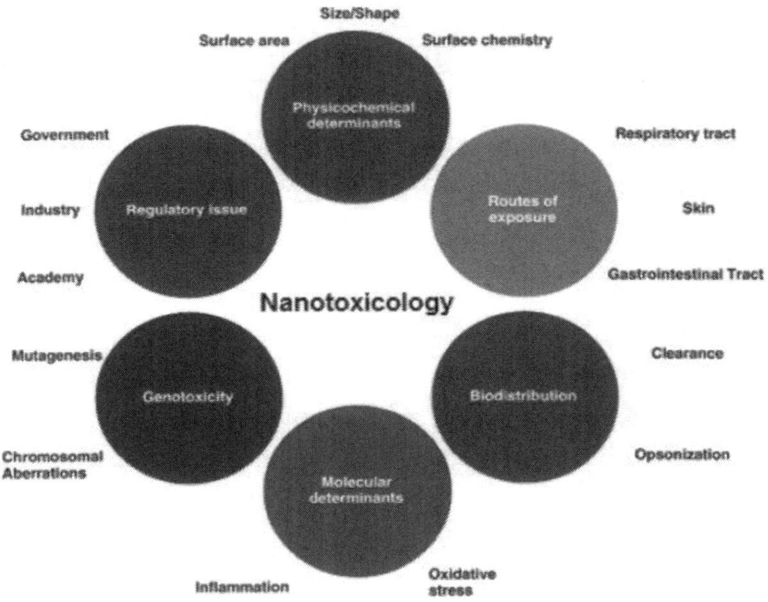

Fig 36. Nanotoxicology

Another challenge is the potential for unintended exposure to nanomaterials in the environment and in the workplace. Due to their small size and unique properties, nanomaterials may be able to penetrate biological barriers, such as the blood-brain barrier, and may accumulate in organs and tissues, potentially leading to chronic exposure and toxicity. As a result, there is a need for better monitoring and regulation of nanomaterials to ensure their safe use.

Overall, nanotoxicology plays a critical role in ensuring the safe and responsible use of nanomaterials. As the field continues to grow, it is important to continue developing standardized methods for toxicity testing and risk assessment, as well as improving our understanding of the potential hazards and risks associated with nanomaterials

17.2 Toxicity of Nanoparticles

Nanoparticles are tiny particles that range in size from 1 to 100 nanometers. Due to their unique physicochemical properties, nanoparticles have found a wide range of applications in various fields, including medicine, electronics, energy, and environmental remediation. However, their small size and large surface area can also result in unique toxicological properties that are not observed in bulk materials.

The toxicity of nanoparticles can vary depending on their size, shape, surface chemistry, and other physicochemical properties. The specific toxicity of nanoparticles refers to their ability to induce adverse effects on living organisms, including humans and the environment, through various mechanisms of action. Here are some examples of specific toxicity of nanoparticles:

a. Reactive oxygen species (ROS) generation: Some nanoparticles, such as titanium dioxide ($TiO2$), zinc oxide (ZnO), and carbon black, have been shown to generate ROS, which can cause oxidative stress and damage to DNA, proteins, and lipids. This can lead to inflammation, cell death, and tissue damage.
b. Inflammation: Nanoparticles can also induce inflammation through various mechanisms, such as activation of the immune system and release of pro-inflammatory cytokines. This can lead to tissue damage and contribute to the development of various diseases, such as asthma, lung cancer, and cardiovascular disease.
c. DNA damage: Some nanoparticles, such as silver nanoparticles, have been shown to induce DNA damage

and mutations, which can contribute to the development of cancer and other genetic diseases.

d. Disruption of cellular processes: Nanoparticles can also disrupt various cellular processes, such as protein synthesis, ion transport, and cell signaling, leading to dysfunction and cell death. For example, some nanoparticles, such as gold nanoparticles, have been shown to disrupt the mitochondrial function, leading to cell death.
e. Neurotoxicity: Some nanoparticles, such as iron oxide nanoparticles and quantum dots, can cross the blood-brain barrier and accumulate in the brain, potentially leading to neurotoxicity. This can result in cognitive impairment, memory loss, and other neurological disorders.
f. Environmental toxicity: Nanoparticles can also have adverse effects on the environment, such as altering the behavior and survival of aquatic organisms and disrupting the food chain. For example, some nanoparticles, such as silver nanoparticles, have been shown to be toxic to aquatic organisms, such as fish and algae.

It is important to note that the toxicity of nanoparticles can also depend on the dose and duration of exposure. Some nanoparticles may be safe at low concentrations but become toxic at higher concentrations or after chronic exposure. In addition, the toxicity of nanoparticles can also depend on the route of exposure, such as inhalation, ingestion, or dermal exposure.

To better understand the specific toxicity of nanoparticles, various in vitro and in vivo toxicity testing methods have been developed. These methods include cell culture assays, animal studies, and environmental toxicity testing. However, the standardization and validation of these methods for nanoparticle toxicity testing are still ongoing.

Overall, the specific toxicity of nanoparticles is an important area of research and can have significant implications for human health and the environment. As the use of nanoparticles continues to grow, it is important to continue studying their toxicological properties and developing safe and responsible practices for their use.

17.3 Safety and Regulatory Issues of Nanoparticles

The use of nanoparticles in various fields, such as medicine, electronics, and energy, has been growing rapidly. However, as with any new technology, there are concerns about the safety and regulatory issues of nanoparticles. Due to their small size and unique physicochemical properties, nanoparticles can potentially pose risks to human health and the environment. Therefore, it is essential to understand and address these concerns to ensure safe and responsible use of nanoparticles.

Safety Issues:

Exposure: One of the primary safety concerns of nanoparticles is exposure. Due to their small size, nanoparticles can easily enter the body through inhalation, ingestion, or skin contact. Once inside the body, nanoparticles can accumulate in various organs, potentially causing adverse effects. Therefore, it is essential to ensure that exposure to nanoparticles is minimized, and safety guidelines are followed to protect workers and the general public.

Toxicity: Nanoparticles can potentially induce adverse effects on human health, such as inflammation, DNA damage, and cell death. The toxicity of nanoparticles can vary depending on their size, shape, surface chemistry, and other physicochemical properties. Therefore, it is essential to study the toxicological properties of nanoparticles and develop safety guidelines to minimize their adverse effects.

Environmental impact: Nanoparticles can also have adverse effects on the environment. For example, nanoparticles can accumulate in soil and water, potentially disrupting the food chain and harming aquatic organisms. Therefore, it is essential to study the environmental impact of nanoparticles and develop safety guidelines to minimize their adverse effects.

Regulatory Issues:

Lack of regulations: There is currently a lack of regulatory oversight of nanoparticles in many countries. This is due to the fact that nanoparticles are relatively new, and regulatory agencies are still developing guidelines and regulations for their safe use. Therefore, it is essential to develop robust regulatory frameworks to ensure the safe use of nanoparticles in various fields.

Lack of standardized testing methods: There is currently a lack of standardized testing methods for the toxicity of nanoparticles. This makes it difficult to compare the toxicity of different nanoparticles and develop safety guidelines. Therefore, it is essential to develop standardized testing methods to evaluate the toxicity of nanoparticles.

Labeling requirements: There is currently a lack of labeling requirements for products containing nanoparticles. This makes it difficult for consumers to identify and avoid products that may contain potentially harmful nanoparticles. Therefore, it is essential to develop labeling requirements for products containing nanoparticles to ensure transparency and inform consumers.

International harmonization: There is currently a lack of international harmonization of regulations for nanoparticles. This can create confusion and barriers to trade, as different countries may have different regulations. Therefore, it is essential to develop international harmonization of regulations for nanoparticles to ensure consistency and facilitate international trade.

Addressing Safety and Regulatory Issues:

Risk assessment: It is essential to conduct risk assessments of nanoparticles to identify potential hazards and develop appropriate safety guidelines. Risk assessments should consider exposure routes, toxicity, and environmental impact.

Standardized testing methods: Developing standardized testing methods for the toxicity of nanoparticles can facilitate the evaluation of the toxicity of different nanoparticles and develop safety guidelines.

Regulatory oversight: Developing robust regulatory oversight of nanoparticles can ensure their safe use in various fields. Regulatory oversight should consider the entire lifecycle of nanoparticles, including manufacturing, use, and disposal.

International cooperation: Developing international cooperation on the regulation of nanoparticles can facilitate the harmonization of regulations and ensure consistency in the regulation of nanoparticles globally.

Transparency and education: Increasing transparency and education on the use of nanoparticles can inform consumers and workers about potential hazards and promote safe and responsible use of nanoparticles.

In conclusion, nanoparticles offer many benefits in various fields. However, it is essential to address the safety and regulatory issues of nanoparticles to ensure their safe and responsible use. This includes conducting risk assessments, developing standardized testing methods, establishing regulatory oversight, promoting international cooperation, and increasing transparency and education on the use of nanoparticles. By addressing these issues, we can ensure the safe and responsible use of nanoparticles, while maximizing their potential benefits and minimizing their potential risks to human health and the environment.

CHALLENGES

- What is nanotoxicology?
- What are the potential hazards associated with exposure to nanoparticles?
- What are the different routes of exposure to nanoparticles?
- How are nanoparticles absorbed and distributed in the body?
- What are the different types of toxicity associated with nanoparticles?
- What are some factors that can influence the toxicity of nanoparticles?
- How can the toxicity of nanoparticles be assessed?
- What are some in vitro and in vivo toxicity testing methods used in nanotoxicology?
- What are the ethical considerations surrounding nanotoxicology research?
- How can nanotoxicology research be used to inform regulation and policy?
- What are the challenges in developing standardized protocols for nanotoxicology testing?
- How can advances in nanotoxicology research be used to improve the safety of nanotechnology?
- What are some potential risks associated with the use of nanotechnology in consumer products?

CHAPTER EIGHTEEN

18.1 Nanomaterials for Food Applications

Nanomaterials have been increasingly used in the food industry due to their unique physicochemical properties, such as high surface area, high reactivity, and tunable functionality. Nanomaterials offer potential benefits in various food applications, such as food packaging, food processing, and food safety. However, there are also concerns about the safety and regulatory issues of using nanomaterials in food.

Nanomaterials in Food Packaging:

Nanomaterials can be used in food packaging to enhance the barrier properties, mechanical strength, and antimicrobial properties of packaging materials. For example, nanocomposite films made from polymer matrices and nanoparticles can provide improved oxygen barrier properties and mechanical strength, thus prolonging the shelf-life of food products. In addition, nanoparticles, such as silver nanoparticles, can be incorporated into packaging materials to provide antimicrobial properties, thus reducing the risk of foodborne illness.

Fig 37. Food packages

Nanomaterials in Food Processing:

Nanomaterials can also be used in food processing to improve the texture, flavor, and nutrient content of food products. For example, nanoemulsions can be used as a delivery system for bioactive compounds, such as vitamins and antioxidants, to enhance their stability and bioavailability in food products. In addition, nanocarriers, such as liposomes and solid lipid nanoparticles, can be used to encapsulate sensitive ingredients, such as enzymes and probiotics, to protect them from degradation during processing and storage.

Nanomaterials in Food Safety:

Nanomaterials can also be used in food safety applications, such as foodborne pathogen detection and removal. For example, nanoparticles, such as quantum dots and gold nanoparticles, can be used as labels for rapid and sensitive detection of foodborne pathogens. In addition, nanoparticles, such as titanium dioxide and zinc oxide, can be used as photocatalysts for the removal of foodborne pathogens in water and air.

Safety and Regulatory Issues:

There are concerns about the safety and regulatory issues of using nanomaterials in food. One of the primary concerns is the potential toxicity of nanomaterials, as they may penetrate cell membranes and interact with biological systems differently than their bulk counterparts. Therefore, it is essential to conduct thorough toxicological studies to evaluate the safety of nanomaterials in food.

Another concern is the lack of regulatory oversight of nanomaterials in food. Currently, there are no specific regulations or labeling requirements for nanomaterials in food in many countries. Therefore, it is essential to develop robust regulatory frameworks to ensure the safe use of nanomaterials in food and inform consumers about the presence of nanomaterials in food products.

In conclusion, nanomaterials offer potential benefits in various food applications, such as food packaging, food processing, and food safety. However, it is essential to address the safety and regulatory issues of using nanomaterials in food to ensure their safe and responsible use. This includes conducting thorough toxicological studies, developing regulatory frameworks, and promoting transparency and education on the use of nanomaterials in food. By addressing these issues, we can maximize the potential benefits of nanomaterials in food while minimizing their potential risks to human health and the environment.

Nanoscale Delivery of Nutrients and Flavorings

Nanotechnology has the potential to revolutionize the delivery of nutrients and flavorings in the food industry. By using nanoscale particles, it is possible to enhance the bioavailability, stability, and sensory properties of nutrients and flavorings in food products. Here are some examples of how nanotechnology can be used for the delivery of nutrients and flavorings:

Nanoemulsions for Nutrient Delivery:

Nanoemulsions are colloidal systems composed of oil, water, and an emulsifier, with droplet sizes ranging from 20-200 nm. Nanoemulsions can be used as delivery systems for lipophilic nutrients, such as vitamins A, D, E, and K, and omega-3 fatty acids. By encapsulating these nutrients in nanoemulsions, their bioavailability and stability can be improved, which can enhance their health benefits.

Encapsulation for Nutrient Protection:

Encapsulation is a process in which a material is surrounded by a coating or matrix, which can protect it from degradation or interact with the surrounding environment. Nanoparticles, such as liposomes, solid lipid nanoparticles, and polymeric nanoparticles, can be used as carriers for the encapsulation of sensitive nutrients, such as probiotics and enzymes. By protecting these nutrients from environmental factors, such as pH and temperature, their activity and functionality can be maintained during processing and storage.

Nanoscale Coatings for Flavorings:

Nanoscale coatings can be used to enhance the sensory properties of food products by providing controlled release of flavorings. For example, nanoparticles, such as cyclodextrins, can encapsulate flavor compounds and slowly release them over time, providing a prolonged flavor experience. Nanoparticles can also be used to mask bitter or unpleasant flavors by coating the taste buds with a bitter-blocking agent.

Nanoparticles for Color Enhancement:

Nanoparticles can also be used to enhance the color of food products. For example, titanium dioxide nanoparticles can be used as a white pigment to improve the appearance of dairy products and confectionery. Iron oxide nanoparticles can also be used as a red pigment to enhance the color of meat products.

However, there are concerns about the safety and regulatory issues of using nanotechnology in food. One of the primary concerns is the potential toxicity of nanomaterials, as they may penetrate cell membranes and interact with biological systems differently than their bulk counterparts. Therefore, it is essential to conduct thorough toxicological studies to evaluate the safety of nanomaterials in food.

Another concern is the lack of regulatory oversight of nanomaterials in food. Currently, there are no specific regulations or labeling requirements for nanomaterials in food in many countries. Therefore, it is essential to develop robust regulatory frameworks to ensure the safe use of nanomaterials in food and inform consumers about the presence of nanomaterials in food products.

In conclusion, nanotechnology has the potential to revolutionize the delivery of nutrients and flavorings in the food industry. By using nanoscale particles, it is possible to enhance the bioavailability, stability, and sensory properties of nutrients and flavorings in food products. However, it is essential to address the safety and regulatory issues of using nanotechnology in food to ensure their safe and responsible use. This includes conducting thorough toxicological studies, developing regulatory frameworks, and promoting transparency and education on the use of nanotechnology in food. By addressing these issues, we can maximize the potential benefits of nanotechnology in food while minimizing their potential risks to human health and the environment.

CHALLENGES

- What are nanomaterials for food packaging, food processing, and nutrient delivery?
- What are the benefits of using nanomaterials in food packaging?
- How can nanomaterials be used to increase the shelf life of food?
- What are some types of nanomaterials used in food packaging?
- What are the challenges associated with incorporating nanomaterials into food packaging?
- What are some concerns regarding the potential release of nanomaterials from food packaging into food?
- What are some applications of nanomaterials in food processing?
- How can nanomaterials be used to improve the texture, flavor, and nutritional content of food?
- What are some challenges associated with incorporating nanomaterials into food processing?
- How can nanomaterials be used for nutrient delivery?
- What are some types of nanomaterials used for nutrient delivery?
- How can nanomaterials be used to enhance the bioavailability of nutrients?
- What are some concerns regarding the safety of nanomaterials used for nutrient delivery?
- What are some potential risks associated with the use of nanomaterials in food packaging, food processing, and nutrient delivery?
- How can the safety of nanomaterials be assessed for use in food applications?

- What are the ethical considerations surrounding the use of nanomaterials in food applications?
- What are some potential environmental impacts of the use of nanomaterials in food applications?
- What is the future of nanomaterials for food packaging, food processing, and nutrient delivery?
- How can advances in synthesis and characterization techniques improve the performance of nanomaterials for food applications?
- How can nanomaterials for food applications be made more accessible and affordable for developing countries?

CHAPTER NINETEEN

19.1 Nanotechnology in Agriculture

Nanotechnology has the potential to revolutionize agriculture by offering innovative solutions for sustainable crop production, soil health, and plant protection. Nanotechnology in agriculture involves the application of nanomaterials and nanodevices to improve plant growth, nutrient uptake, disease resistance, and pest control. Here are some examples of how nanotechnology can be used in agriculture:

19.2 Nanofertilizers for Improved Nutrient Uptake

Nanofertilizers are nanoscale particles that can deliver nutrients to plants in a more efficient and targeted way. By encapsulating nutrients, such as nitrogen, phosphorus, and potassium, in nanoscale particles, their availability and uptake by plants can be improved, reducing the amount of fertilizer needed and minimizing environmental pollution. Nanofertilizers can also be designed to release nutrients slowly over time, improving their efficiency and reducing nutrient loss.

19.3 Nanosensors for Soil Health Monitoring

Nanosensors can be used to monitor soil health by measuring soil properties such as pH, moisture, and nutrient levels. By providing real-time data on soil conditions, farmers can optimize crop management practices, such as irrigation and fertilizer application, to improve crop yield and quality. Nanosensors can also be used to detect soil contaminants, such as heavy metals and pesticides, which can help prevent their accumulation in food crops.

19.4 Nanopesticides for Improved Pest Control

Nanopesticides are nanoscale particles that can be used to control pests, such as insects and fungi, in a more efficient and targeted way.

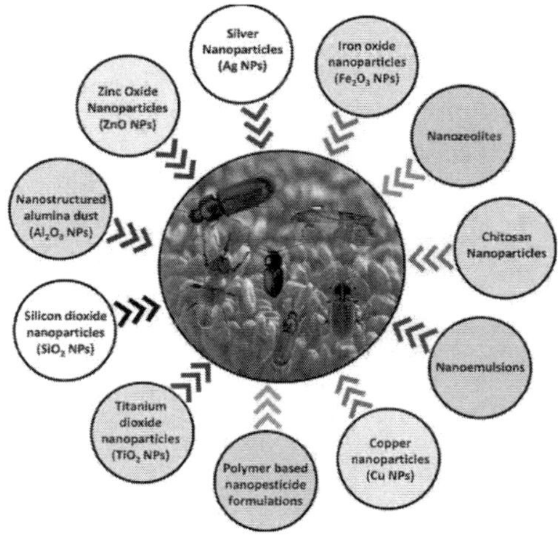

Fig 38. Nanomaterials in pest control

By encapsulating pesticides in nanoscale particles, their delivery to pests can be improved, reducing the amount of pesticide needed and minimizing environmental pollution.Nanopesticides can also be designed to release pesticides slowly over time, improving their efficiency and reducing pesticide drift.

19.5 Nanocarriers for Plant Protection

Nanocarriers can be used to deliver bioactive compounds, such as plant hormones and proteins, to plants for disease resistance and stress tolerance. By encapsulating these compounds in nanoscale particles, their stability and bioavailability can be improved, enhancing their effectiveness. Nanocarriers can also be designed to target specific plant tissues, such as roots or leaves, for optimal delivery.

In conclusion, nanotechnology has the potential to revolutionize agriculture by offering innovative solutions for sustainable crop production, soil health, and plant protection. By using nanoscale particles and devices, it is possible to improve nutrient uptake, pest control, and plant protection while reducing environmental pollution. However, it is essential to address the safety and regulatory issues of using nanotechnology in agriculture to ensure their safe and responsible use. This includes conducting thorough toxicological studies, developing regulatory frameworks, and promoting transparency and education on the use of nanotechnology in agriculture. By addressing these issues, we can maximize the potential benefits of nanotechnology in agriculture while minimizing their potential risks to human health and the environment.

19.6 Nano-based Sensors for Crop Monitoring

Nano-based sensors have the potential to revolutionize crop monitoring by providing real-time data on crop growth and health. These sensors can detect changes in temperature, humidity, nutrient levels, and disease outbreaks, allowing farmers to optimize crop management practices and improve yields.

Some examples of nano-based sensors for crop monitoring include:

Carbon nanotube sensors:

Carbon nanotubes are nanoscale tubes made of carbon atoms that can detect changes in temperature and humidity in crops. These sensors can be embedded in soil or plant tissue and connected to a wireless network to transmit data to farmers in real-time.

Nanoparticle-based sensors:

Nanoparticle-based sensors can detect changes in nutrient levels, such as nitrogen, phosphorus, and potassium, in soil and plant tissue. These sensors can also detect heavy metals and other contaminants that can affect crop growth and health.

Quantum dot sensors:

Quantum dots are nanoscale semiconductors that can emit light at different wavelengths when exposed to certain chemicals or substances. These sensors can detect changes in plant hormones and other biochemicals that indicate stress or disease in crops.

Nanopore sensors:

Nanopore sensors can detect changes in DNA and RNA sequences in plant tissue, which can indicate the presence of pathogens or mutations that affect crop growth and health.

Overall, nano-based sensors have the potential to transform agriculture by providing farmers with real-time data on crop growth and health. By monitoring crops at the nanoscale level, farmers can optimize crop management practices and improve yields while minimizing environmental pollution.

However, there are concerns about the safety and regulatory issues of using nanotechnology in agriculture. One of the primary concerns is the potential toxicity of nanomaterials, as they may interact with living organisms differently than their bulk counterparts. Therefore, it is essential to conduct thorough toxicological studies to evaluate the safety of nanomaterials in agriculture.

Another concern is the lack of regulatory oversight of nanomaterials in agriculture. Currently, there are no specific regulations or labeling requirements for nanomaterials in agricultural products in many countries. Therefore, it is essential to develop robust regulatory frameworks to ensure the safe use of nanomaterials in agriculture and inform consumers about the presence of nanomaterials in food products.

CHALLENGES

- What are nanomaterials for fertilizers, crop protection, and crop monitoring?
- How can nanomaterials be used in fertilizers to increase crop yields?
- What are some types of nanomaterials used in fertilizers?
- What are the challenges associated with incorporating nanomaterials into fertilizers?
- What are some concerns regarding the safety of nanomaterials used in fertilizers?
- How can nanomaterials be used in crop protection to prevent pest infestations?
- What are some types of nanomaterials used in crop protection?
- What are the challenges associated with incorporating nanomaterials into crop protection products?
- What are some concerns regarding the safety of nanomaterials used in crop protection?
- How can nanomaterials be used in crop monitoring to assess plant health and detect environmental stresses?
- What are some types of nanomaterials used in crop monitoring?
- What are the challenges associated with incorporating nanomaterials into crop monitoring systems?
- What are some potential environmental impacts of the use of nanomaterials in agriculture?
- How can the safety of nanomaterials be assessed for use in agricultural applications?
- What are the ethical considerations surrounding the use of nanomaterials in agriculture?
- What are some potential risks associated with the use of nanomaterials in agriculture?

- How can advances in synthesis and characterization techniques improve the performance of nanomaterials for agricultural applications?
- What is the future of nanomaterials for fertilizers, crop protection, and crop monitoring?
- How can nanomaterials for agricultural applications be made more accessible and affordable for small-scale farmers?
- What are some potential benefits of using nanomaterials in agriculture for sustainable farming practices?

CHAPTER TWENTY

20.1 Nanotechnology in Textiles

Nanotechnology has emerged as a promising field for the development of advanced textiles with enhanced properties and functionalities. By manipulating materials at the nanoscale level, it is possible to create textiles with improved durability, water and stain resistance, and UV protection, among other desirable features.

Nanoparticles, nanofibers, and nanocomposites are some of the nanomaterials that can be used to enhance the properties of textiles. For instance, nanoparticles of metals such as silver and copper can be incorporated into textiles to create antimicrobial fabrics that can kill bacteria and fungi. Similarly, nanofibers made of materials such as carbon and cellulose can be used to create textiles with enhanced strength and durability.

One of the most common applications of nanotechnology in textiles is the development of water and stain-resistant fabrics. Nanoparticles such as titanium dioxide and silicon dioxide can be applied to fabrics to create a hydrophobic surface that repels water and other liquids. This can make textiles more resistant to staining and extend their lifespan.

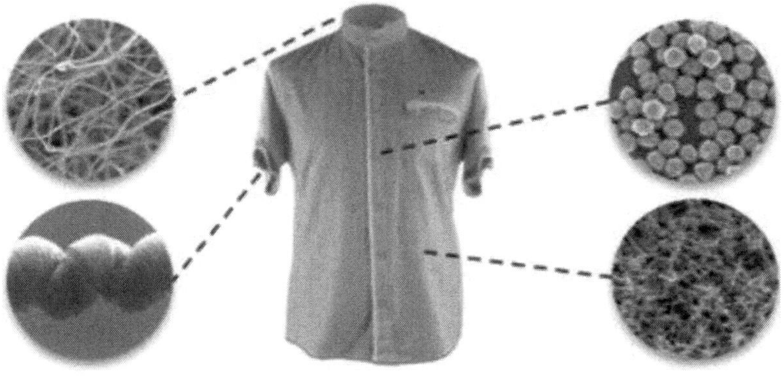

Fig 39. Nanomaterials in textiles

Nanotechnology can also be used to enhance the UV protection properties of textiles. Nanoparticles such as zinc oxide and titanium dioxide can be incorporated into textiles to create fabrics that can protect against harmful UV radiation from the sun.

In addition to these applications, nanotechnology can be used to develop smart textiles that can sense and respond to changes in the environment. For instance, nanosensors can be embedded into textiles to monitor vital signs or detect changes in temperature, humidity, or other environmental factors.

Overall, nanotechnology has the potential to revolutionize the textile industry by creating advanced fabrics with enhanced properties and functionalities. These textiles can offer improved performance, durability, and comfort, and can find applications in a wide range of industries, from fashion and sportswear to healthcare and aerospace.

Nano-based Fabrics for Performance and Comfort

Nanotechnology has opened up new possibilities for the development of advanced fabrics with enhanced performance and comfort features. By manipulating materials at the nanoscale level, it is possible to create textiles with improved properties such as breathability, moisture management, and thermal regulation.

One of the most significant applications of nanotechnology in fabrics is the development of moisture-wicking and breathable textiles. Nanoparticles such as titanium dioxide and silica can be incorporated into fabrics to create a hydrophilic surface that wicks away moisture from the skin and enhances breathability. This can help to keep the wearer dry and comfortable, particularly during physical activity or in hot and humid conditions.

Nanotechnology can also be used to develop fabrics with enhanced thermal regulation properties. For instance, phase-change materials (PCMs) can be incorporated into textiles to absorb and release heat as needed, helping to regulate body temperature and improve comfort. Nanofibers and nanocomposites can also be used to create fabrics with improved insulation properties, keeping the wearer warm in cold weather and cool in hot weather.

In addition to these performance features, nanotechnology can also be used to develop fabrics with enhanced durability and resistance to wear and tear. Nanoparticles such as silver and copper can be incorporated into fabrics to create antimicrobial properties that can kill bacteria and fungi, reducing odor and extending the life of the fabric. Nanofibers can also be used to create fabrics with improved tensile strength and abrasion resistance, making them more durable and long-lasting.

Overall, nanotechnology has the potential to transform the textile industry by creating advanced fabrics with enhanced properties and functionalities. These textiles can offer improved performance, comfort, and durability, and can find applications in a wide range of industries, from sportswear and outdoor gear to healthcare and aerospace.

CHALLENGES

What are nanomaterials for textile applications?

How can nanomaterials be used in textiles to enhance their performance?

What are some types of nanomaterials used in textiles?

What are the challenges associated with incorporating nanomaterials into textiles?

What are some concerns regarding the safety of nanomaterials used in textiles?

How can nanomaterials be used to improve the durability and strength of textiles?

How can nanomaterials be used to enhance the functionality of textiles, such as water resistance, stain resistance, and UV protection?

How can nanomaterials be used to create smart textiles with sensing and self-cleaning properties?

How can nanomaterials be used to improve the comfort and breathability of textiles?

What are some potential environmental impacts of the use of nanomaterials in textiles?

How can the safety of nanomaterials be assessed for use in textile applications?

What are the ethical considerations surrounding the use of nanomaterials in textiles?

What are some potential risks associated with the use of nanomaterials in textiles?

How can advances in synthesis and characterization techniques improve the performance of nanomaterials for textile applications?

What is the future of nanomaterials for textile applications and how can they be made more sustainable and affordable?

CHAPTER TWENTY-ONE

21.1 Nanodevices

Nanodevices refer to electronic or mechanical devices with at least one dimension measured in nanometers (nm), typically ranging from 1-100 nm. Due to their small size, they exhibit unique physical and chemical properties that are not found in larger structures. This has opened up numerous possibilities in a variety of fields, including electronics, medicine, energy, and materials science.

One of the most promising applications of nanodevices is in the field of electronics. The continued miniaturization of electronic components has led to the development of transistors, the fundamental building block of all modern electronic devices. The current state-of-the-art transistor technology is based on complementary metal-oxide-semiconductor (CMOS) technology, which uses silicon as the base material. However, as transistors continue to shrink, they face several challenges such as increased leakage current and higher power consumption, which limits their speed and efficiency.

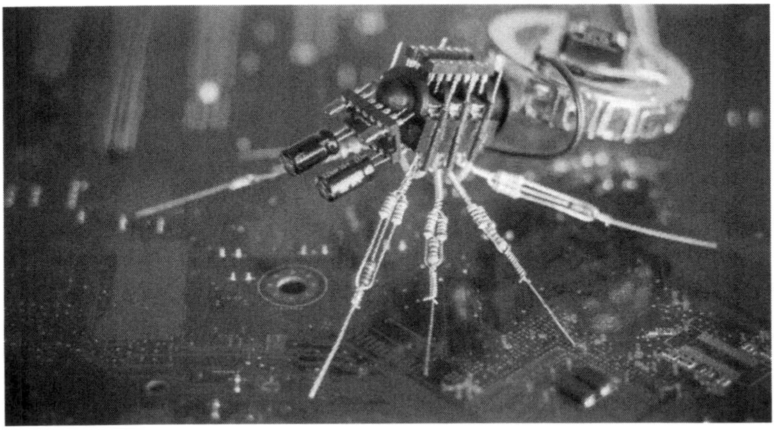

Fig 40. Nanodevices

Nanodevices offer a potential solution to these problems. For example, carbon nanotubes (CNTs) and graphene are two promising materials for constructing nanoscale transistors. CNTs are hollow tubes made of carbon atoms and have unique electrical and mechanical properties. They have been shown to outperform traditional silicon transistors in terms of speed and power consumption. Graphene, on the other hand, is a single layer of carbon atoms arranged in a hexagonal lattice. It has excellent electrical conductivity and can be used to make highly efficient and flexible electronic devices.

Nanodevices also have potential applications in the field of medicine. One area of research is the development of nanorobots, which are tiny machines designed to perform specific tasks within the body, such as delivering drugs or repairing damaged tissues. These devices can be constructed using materials such as DNA, proteins, or nanoparticles, and can be programmed to respond to specific signals within the body, such as changes in pH or the presence of certain molecules.

Another promising application of nanodevices in medicine is in the field of diagnostic imaging. Nanoparticles can be designed to selectively bind to specific cells or tissues, allowing them to be detected using imaging techniques such as magnetic resonance imaging (MRI) or positron emission tomography (PET). This can improve the accuracy of cancer detection and help monitor the effectiveness of treatments

Nanodevices can also play a role in addressing global energy challenges. For example, they can be used to improve the efficiency of solar cells by increasing their surface area and reducing the amount of material required. Nanodevices can also be used to develop more efficient batteries and fuel cells, which could help reduce our dependence on fossil fuels.

In addition to these applications, nanodevices have the potential to revolutionize materials science. By designing materials at the nanoscale, researchers can tailor their properties to suit specific applications. For example, nanoparticles can be used to enhance the strength and durability of materials, or to create materials with new optical or magnetic properties.

Despite the many potential applications of nanodevices, there are also several challenges that need to be addressed. One of the main challenges is the difficulty in controlling the properties of nanoscale structures. At this scale, the behavior of materials can be unpredictable, and small changes in structure or composition can have a significant impact on their properties. Additionally, the production of nanodevices can be expensive and time-consuming, and there is still much research to be done to fully understand their potential risks and benefits.

21.2 Nanobots

Nanobots, also known as nanorobots or nanomachines, are small machines or robots that are built on a nanoscale level, typically in the range of 0.1 to 10 micrometers. They are designed to perform specific tasks, such as delivering drugs to targeted cells, repairing damaged tissues, or cleaning up pollutants.

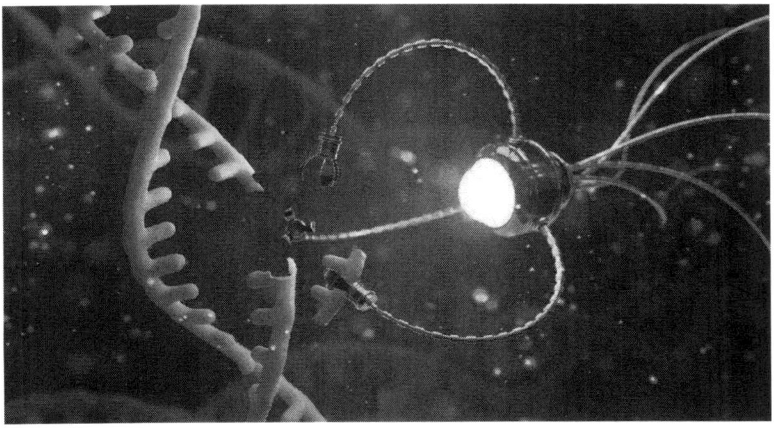

Fig 41. Nanobots

The development of nanobots is a rapidly advancing field that has the potential to revolutionize many aspects of our lives. There are many potential applications for their use in medicine, industry, and environmental cleanup. However, there are also concerns about the potential risks associated with their use, including unintended consequences and the possibility of unintended harm to living organisms.

One of the most promising applications of nanobots is in the field of medicine. Nanobots can be designed to target specific cells or tissues within the body, allowing for more precise drug delivery and reducing the risk of side effects. For example, they could be used to deliver chemotherapy drugs directly to cancer cells, minimizing damage to healthy cells. They could also be used to repair damaged tissues or organs by delivering growth factors or other biological agents.

Another potential application of nanobots is in the field of nanoelectronics. Nanobots could be used to create new electronic devices that are much smaller and more efficient than current technology. They could also be used to create new materials with unique properties, such as superconductivity or extreme strength.

Nanobots could also be used for environmental cleanup. They could be designed to target and break down pollutants, such as oil spills or plastic waste, into harmless substances. This could have a significant impact on the health of our planet and help to reduce the impact of human activity on the environment.

Despite the many potential benefits of nanobots, there are also concerns about their safety and potential unintended consequences. For example, there is a risk that nanobots could escape from their intended target and cause unintended harm to living organisms. There is also a risk that they could be used for nefarious purposes, such as espionage or terrorism.

and potential unintended consequences. For example, there is a risk that nanobots could escape from their intended target and cause unintended harm to living organisms. There is also a risk that they could be used for nefarious purposes, such as espionage or terrorism.

To mitigate these risks, researchers are working to develop safe and effective nanobots that are designed with strict safety protocols in mind. This includes designing nanobots that are biodegradable, so that they break down into harmless substances after their intended use. It also includes developing strict regulations and oversight to ensure that nanobots are used only for their intended purpose and do not cause harm to living organisms.

In conclusion, nanobots are a rapidly advancing field that has the potential to revolutionize many aspects of our lives, from medicine to electronics to environmental cleanup. While there are concerns about their safety and potential unintended consequences, researchers are working to develop safe and effective nanobots that can be used for the benefit of society.

21.2 Artificial Intelligence and Nanobots

Artificial intelligence (AI) and nanobots are two rapidly advancing fields of technology that have the potential to work together to create new and innovative solutions to many problems.

One potential application of AI and nanobots is in the field of medicine. AI can be used to analyze large amounts of medical data and identify patterns that may be difficult for humans to see. Nanobots, on the other hand, can be designed to deliver drugs or other treatments to specific cells or tissues within the body. By combining AI with nanobots, it may be possible to create personalized medicine that is tailored to the individual patient's unique needs.

Another potential application of AI and nanobots is in the field of environmental cleanup. AI can be used to analyze large amounts of data about environmental pollutants and identify areas where cleanup efforts would be most effective. Nanobots, on the other hand, can be designed to target and break down pollutants into harmless substances. By combining AI with nanobots, it may be possible to create a more efficient and effective approach to environmental cleanup.

AI can also be used to control and direct the movements of nanobots. For example, AI algorithms could be used to control a swarm of nanobots to clean up a polluted waterway or to repair damaged tissues within the body. This could lead to more precise and efficient outcomes than would be possible with human intervention alone.

However, there are also concerns about the potential risks associated with the use of AI and nanobots. For example, there is a risk that AI algorithms could malfunction and direct nanobots to unintended targets, causing unintended harm to living organisms. There is also a risk that nanobots could be used for malicious purposes, such as espionage or terrorism.

To mitigate these risks, researchers are working to develop safe and effective AI and nanobots that are designed with strict safety protocols in mind. This includes designing nanobots that are biodegradable, so that they break down into harmless substances after their intended use. It also includes developing strict regulations and oversight to ensure that AI and nanobots are used only for their intended purpose and do not cause harm to living organisms.

In conclusion, AI and nanobots are two rapidly advancing fields of technology that have the potential to work together to create new and innovative solutions to many problems. While there are concerns about the potential risks associated with their use, researchers are working to develop safe and effective AI and nanobots that can be used for the benefit of society.

Nanochemistry has opened up new opportunities for the development of advanced materials with unique and enhanced properties. However, there are several challenges that must be addressed to ensure the safe and effective use of nanomaterials in various applications.

One of the main challenges is the potential toxicity of nanomaterials. The small size and high surface area of nanoparticles can lead to increased reactivity and potential harm to living organisms, including humans. Therefore, it is essential to understand the potential toxicity of nanomaterials and develop appropriate safety guidelines for their use.

Another challenge is the scalability of nanomaterials production. While many nanomaterials can be synthesized in the lab, scaling up their production to industrial levels can be challenging. This is due to the complexity of the production process, the need for specialized equipment, and the high cost of materials. Developing efficient and cost-effective methods for large-scale production of nanomaterials is crucial for their widespread use.

The characterization of nanomaterials is also a challenge. The unique properties of nanoparticles require specialized characterization techniques to fully understand their physical, chemical, and biological properties. However, many traditional characterization techniques are not suitable for nanomaterials, and new methods must be developed to fully understand the properties and behavior of these materials.

The regulation and standardization of nanomaterials is another important challenge. There are currently no universal regulations governing the production, use, and disposal of nanomaterials, and there is a lack of standardization in testing and characterization methods. This can lead to inconsistencies in the quality and safety of nanomaterials and makes it difficult to ensure their safe use.

Despite these challenges, the future of nanochemistry is promising. With continued research and development, it is possible to overcome these challenges and unlock the full potential of nanomaterials in various applications.

In the field of energy, nanomaterials have the potential to revolutionize energy storage and conversion. Nanomaterials can improve the performance and efficiency of batteries, fuel cells, and solar cells, making them more practical and cost-effective.

In healthcare, nanomaterials can be used to develop advanced drug delivery systems that can target specific cells and tissues in the body. This can lead to more effective treatments with fewer side effects. Nanomaterials can also be used to develop advanced medical devices and diagnostic tools.

In the field of electronics, nanomaterials can be used to develop faster and more efficient computer chips, sensors, and other electronic devices. Nanomaterials can also be used to develop flexible and transparent electronics, opening up new possibilities for wearable technology and other applications.

Overall, the future of nanochemistry is bright, with many exciting opportunities for the development of advanced materials with unique and enhanced properties. Addressing the challenges of toxicity, scalability, characterization, and regulation is crucial for the safe and effective use of nanomaterials in various applications. With continued research and development, nanomaterials have the potential to revolutionize many industries and improve the quality of life for people around the world.

CHALLENGES

- What are nanodevices and nanobots?
- How are nanodevices and nanobots different from other types of devices and robots?
- What are the potential applications of nanodevices and nanobots?
- What are some types of nanodevices and nanobots currently being developed?
- How are nanodevices and nanobots designed and synthesized?
- What are some challenges associated with designing and synthesizing nanodevices and nanobots?
- What are some methods of characterizing and testing the performance of nanodevices and nanobots?
- What are some ethical considerations associated with the development and use of nanodevices and nanobots?
- What are some potential risks associated with the use of nanodevices and nanobots?
- How can the safety of nanodevices and nanobots be assessed?
- What are some potential benefits of using nanodevices and nanobots in medicine?
- How can nanodevices and nanobots be used for drug delivery?
- How can nanodevices and nanobots be used for diagnostics?
- How can nanodevices and nanobots be used for targeted therapy?

Printed in Great Britain
by Amazon

55368325R00117